P9-CKX-327
3 1404 00744 6427

PHASE CHANGE

PHASE CHANGE

The Computer Revolution in Science and Mathematics

DOUGLAS S. ROBERTSON

2003

OXFORD
UNIVERSITY PRESS

Oxford New York
Auckland Bangkok Buenos Aires Cape Town Chennai
Dar es Salaam Delhi Hong Kong Istanbul Karachi Kolkata
Kuala Lumpur Madrid Melbourne Mexico City Mumbai Nairobi
São Paulo Shanghai Taipei Tokyo Toronto

Published by Oxford University Press, Inc.,
198 Madison Avenue, New York, New York 10016

www.oup.com

Library of Cognress Cataloging-in-Publication Data
Robertson, Douglas S.
Phase change : the computer revolution in science
and mathematics / by Douglas S. Robertson.
p. cm.
Includes bibliographical references and index.
ISBN 0-19-515748-6
1. Science—Philosophy. 2. Paradigm (Theory of knowledge)
3. Science—Data processing. 4. Mathematics—Data processing.
I. Title.
Q175.R5224 2003
501—dc21 2002030312

1 2 3 4 5 6 7 8 9

Printed in the United States of America
on acid-free paper

To Joan

PREFACE

This book grew out of a long fascination with the extraordinary power of computer technology, starting with the early punched-paper-tape computers of the 1960s and the pocket calculators of the 1970s, through to the latest microchip technologies and even molecular circuitry in the twenty-first century. Computer technology has made it possible to do many things that could not be done without it. However the main theme of this book is not that computers enable us to do things but rather that computers enable us to see and understand things that could not be seen without them. This newly expanded ability to see and understand things is causing revolutions throughout the sciences and mathematics.

My previous book, *The New Renaissance* (1998), focused on the impact of computers on civilization as a whole rather than on just science and mathematics. That book introduced and explored the concept that civilization is information-limited. The information limits that restrict the development of civilizations can be specified and studied quantitatively, and the first important insight that comes from a quantitative study of information limits is that the revolution generated by computer technology is not the first information revolution in history. It is the fourth.

The first of the three information explosions that preceded the computer revolution is the one that followed the invention of language, the second followed the invention of writing, and the third followed the invention of printing. Language, writing, and printing were the critical information-handling techniques in the precomputer era. And each of these earlier information explosions caused a revolution or major transformation in the overall structure of human society. The invention of language marked the beginning of the human race itself; the invention of writing marked the beginning of classical civilization; and the invention of printing marked the beginning of modern civilization. It is no accident that each of these major transformations in human society is associated with an information explosion; this simply reflects the fact that civilization generally is information-limited. A major increase in our ability to produce, store, and distribute information is therefore the proximate cause of each of the most important revolutions in the history of civilization, and the computer revolution is generating the next step in this sequence, the fourth and by far the largest information explosion in history.

In order to understand the effects of each of these information explosions quantitatively, we have to know how to quantify (measure) information so that the magnitudes of each of these four information explosions can be calculated and compared. Then we have to determine the quantitative information requirements of different levels of civilization in order to understand the impact of an information explosion quantitatively. In my previous book I developed preliminary estimates for all these quantities. I was not arguing that certain quantities of information would force the development of certain types of civilization, but I was arguing the converse: that lack of information-production capacity would forbid the development of certain types of civilization. A continental-scale democracy, for example, is difficult to impossible to achieve without the information-production capability of the printing press. After we have introduced and discussed some elementary concepts from information theory, we will revisit some of these topics in the concluding chapter with a discussion of possible ways to test these ideas. We will identify civilizations that were very similar in many ways until one of them developed quantitative information-production capabilities that the other could not or did not match. Their subsequent development provides interesting insights into the effects of information limits on the development of civilization.

It was clear to me when I wrote my previous book that both science and mathematics were also information-limited (in addition to civilization as a whole). Although this was discussed briefly in chapter 3 of the earlier book, a great deal more can be done with this topic. Both science and mathematics have much more organization and structure than does the rest of civilization, and this structure can be exploited to explore the effects of information explosions in science and mathematics in greater detail than I had attempted before. Chapter 1 will define some of this structure in terms of Thomas Kuhn's famous concept of a paradigm shift, in addition to the concept of a phase change that is central to this book.

The discussions in chapters 2–7 will describe examples of both phase changes and paradigm shifts in a variety of scientific disciplines. These examples are not intended to present a comprehensive picture of any of the various scientific disciplines discussed, still less a complete picture of their history and development. They are intended to be illustrative rather than exhaustive. My intention is to try to be accurate, concise, clear, and not to mislead. "Not misleading" is often a stricter standard than merely accurate. The reader should be warned (as experts in the fields will already be aware) that the subjects I discuss are generally more complex than the picture I have presented. The discussion in this book should provide a useful point to begin exploring those complexities more fully for those who wish to do so. To the extent that I have succeeded in being clear and not misleading, readers should find that the additional complexities that are encountered in their further study will reinforce rather than contradict the conclusions developed here.

Mathematics and the sciences provide a vast cornucopia of examples that I could have selected to illustrate phase changes and paradigm shifts. The selections here naturally reflect my own particular interests and predilections. No two writers would have selected the same set of examples, even from the same fields—there is much more to the earth sciences than probing the internal structure of the Earth, for example, and much more to biology than studies of genomes and proteomes at the molecular level. If I have neglected anyone's particular favorite area of research, I can only apologize in advance. The variety of choices that were available to me is abundant evidence of the incredible richness and diversity of ongoing research in science and mathematics today.

I have attempted to write this book at a level that would be accessible to a talented and motivated undergraduate, but beyond that I have made no special effort to handle each topic to a uniform depth. Again, the variations in the depth of the topics presented here reflect my own interests as well as my judgment as to what might be of most interest to the reading audience.

The order that I have selected to present these topics may appear to be a bit eccentric. I have started with astronomy in chapter 2 because the phase changes there are particularly dramatic and easy to understand, and because I believe that there is broad general interest in astronomy; I hope that many readers will therefore find this to be a congenial starting point. I have followed with a discussion of biological sciences because the invention of the microscope is such an exact parallel to the invention of the telescope. The next topics of physics, mathematics, Earth sciences, and meteorology could have been presented in any order and may be read in any order if the reader is so inclined or has some particular interest in any of these areas. In addition, the discussion in each chapter is not strictly chronological but rather is organized by topic. For example, the chronology of the development of spectroscopy in astronomy overlaps the development of photography, but the two subjects are developed and discussed separately.

The last two chapters take a different tack, developing a theoretical and philosophical framework that will facilitate further exploration of these topics. Chapter 8 will explore some of the deeper philosophical implications of information revolutions by outlining the novel perspective that modern information theory provides into the structure and significance of both phase changes and paradigm shifts. Information theory will be found to provide deep insights into the overall philosophy of science, insights that were unavailable at any time prior to the major developments in information theory that were made in the middle to late decades of the twentieth century. These insights also have major implications for other disciplines outside the sciences and mathematics. The concluding chapter of this book will explore the broader implications of phase changes and information revolutions in the development of civilization.

Throughout the book I will be concerned with quantitative intercomparisons of various revolutions or phase changes in the sciences and mathematics. To describe these comparisons quantitatively I will frequently use the term "order of magnitude." This is a technical term widely used for quantitative estimates in scientific work; it means simply a factor of 10. Thus, if something is ten times larger than another, it is one order of magnitude larger. A factor of 100 is two orders of magnitude (10×10), and a million is six orders of magnitude ($10 \times 10 \times 10 \times 10 \times 10 \times 10$). This exponential scale is convenient for expressing very large differences and it will find repeated uses here.

I am deeply indebted to many colleagues at the University of Colorado and elsewhere for the time and effort they so generously gave to provide invaluable comments and criticism. I would particularly like to thank John Bally, Ben Balsley, Tom Chase, John Cooper, James Faller, Henry Fliegel, Allan Franklin, Rod Frehlich, Alexander Goetz, Michael Grant, Susanna Gross, Kellie Hazell, Ruth Helm, Bruce Jakosky, Michael Jones, Carl Kisslinger, Jorges de S. Martins, Andrew Moore, Norman Pace, Robert Reasenberg, John Rundle, Henry Throop and Steven Weinberg. I am grateful to Marilee DeGoede for editorial advice. The information and insight that these talented and knowledgeable individuals have provided have produced significant improvements to this text. Any errors that remain are, of course, entirely my own responsibility.

The exploration of the impacts of computer technology on science and mathematics in the course of writing this book has been an incredible adventure, a journey marked by both wonder and excitement. I hope that readers will find it as thrilling and fascinating as I did.

CONTENTS

PHASE CHANGE

1

INTRODUCTION

Phase Changes in Science

Bliss was it in that dawn to be alive.

—Wordsworth

Computers are clearly having an enormous impact on all aspects of human society, changing the way we live, work and play, but nowhere is the impact of this technology more important and more far-reaching than in the sciences and mathematics. Researchers today are able to exploit the awesome power of computer technology to accomplish things that their predecessors in the precomputer age could not even dream about: Orbiting telescopes and planetary space probes are expanding our knowledge of the universe. Physicists are exploring 11 dimensions of space-time. Geologists are probing the details of the deep structure of the Earth. Biologists are exploring the human genome and are comparing it to the genomes of other animals and plants and even bacteria to unravel the functions and relationships of genes. Mathematicians are exploring vast new fields including nonlinear dynamics and information theory. And these are not just isolated examples. Major discoveries and conceptual breakthroughs are being made at rates that were inconceivable only a few years ago. This explosion of progress in science and mathematics is not only occurring at unprecedented rates, it is even more remarkable for its extraordinary breadth. It is taking place simultaneously in virtually every area of research. The explosive growth of computer technology has transformed every field of research in ways that are fundamental and important.

But it can be very difficult to comprehend a revolution when you are standing right in the middle of it. To get a better perspective on the computer revolution we need to step back, to examine it from a more distant vantage point. We need to explore and understand its historical parallels. Earlier revolutions in science and mathematics had effects that were very similar to the effects of modern computer technology today, although these earlier revolutions occurred on a much smaller scale.

It is clear that the progress of science and mathematics has never proceeded with uniform speed in the past. Instead, advances have occurred in a sequence of major leaps that were followed by periods of relative quiescence, and these major leaps never occurred simultaneously across every discipline as is happening today. Thomas Kuhn tried to explain the major historical transformations in science and mathematics in terms of a phenomenon that he called a "paradigm shift," but, as we shall see, something else is involved in addition to Kuhn's paradigm shifts. This additional component of scientific revolutions is something that we will call a "phase change," and we will argue that these phase changes are largely responsible for generating Kuhn's paradigm shifts.

Understanding these factors that control the development of science and mathematics is critical to understanding the future progress of science and mathematics, which is a matter that is of intense interest to almost everyone concerned with the subject. David Hilbert expressed this universal interest in his address to the International Congress of Mathematicians in 1900 (quoted in Gray, 2000, p. 240):

> Who among us would not be glad to lift the veil behind which the future lies hidden; to cast a glance at the next advances of our science and at the secrets of its development during future centuries?

While it is never easy to forecast the future course of any complex system, projecting the future of science and mathematics is especially difficult because basic research involves, by definition, an exploration of things that are presently unknown. Nevertheless, a careful examination of the historical development of science and mathematics can give us important insights into its future development. In forecasting the behavior of any system it is common to use models that are smooth and well behaved, typically models that are linear or exponential. But to project the future of science and technology we will need to work with models whose behavior is more drastic than linear or exponential functions. The model that will be found to provide perhaps the best match to the observed changes in the history of science and mathematics is the phenomenon that physicists call a phase change.

A phase change may seem to be an exotic and esoteric concept, but it is actually as familiar as a pitcher full of water and ice cubes. The change from liquid water to ice is a classic example of a phase change. Indeed, the freezing of water is so very familiar that it is easy to overlook how exceedingly strange a phenomenon it really is. Water behaves in an entirely different manner than melted glass, for example. If you cool a batch of melted window glass instead of water, the glass will become thicker and thicker—physicists would say "more viscous"—as the temperature drops, and no phase change occurs. (Viscosity is a technical term that describes a fluid's resistance to flow. Liquid water has a low viscosity, molasses has a higher viscosity.) When water is cooled it behaves entirely differently than glass. It does not become gradually more and more viscous. Indeed, the viscosity of water changes only slightly as the water cools toward the freezing point. But when water reaches zero degrees Celsius, it undergoes a phase change and suddenly becomes solid. In other words, as the temperature drops by a large amount nothing much happens to the viscosity. Then, at a critical temperature, the viscosity suddenly becomes effectively infinite.[1]

It may seem a bit of a stretch to use a model as simple as the freezing of water to understand the behavior of much more complicated systems such as research in science and mathematics. Yet there are two reasons why this is a reasonable model to use. In the first place, as we shall see, there are many events in the history of science and mathematics that have the critical property that is identified with phase changes, that the previous behavior of a system does not give a clue to the future behavior of the system.

The second reason is based on the underlying mathematics of phase changes. The theory of phase changes is a fascinating and active branch of modern chaos theory; see the discussion in Gunton et al. (1973) for an introduction. A full treatment of the theory of phase changes in physical systems is beyond the scope of this book, but one fundamental fact from the theory of phase changes is important for the discussion here: The underlying dynamics of any system that undergoes phase changes is not linear. Linear systems never exhibit phase changes. (For our purposes here the word "linear" simply means that there is some sense in which, if you plot or graph the behavior of the system while you vary some key parameter that controls the system, the resulting graph will be a straight line. This will be discussed in more detail in chapter 4.) This does not mean that phase changes will be found in all possible nonlinear systems, but only that phase changes are a common, almost a characteristic feature of many nonlinear systems. And the behavior of systems of scientific and mathematical research is certainly not linear. Rather, these systems are characterized by significant nonlinear phenomena, including such things as explosive growth and feedback loops. It is therefore reasonable and plausible to expect scientific research to exhibit behaviors that can be modeled as phase changes, even though the underlying nonlinear dynamics of scientific research may be too complex to understand and model in detail. Still, the concept of a phase change as used here should probably be considered more as an analogy than a technically precise definition.

At risk of pushing this analogy too far, there is another aspect of a phase change that may be relevant here. In order for a phase change to occur, a system must be in a critical state. If you cool water from 21 degrees Celsius to 19 degrees Celsius, for example, you do not get a phase change. But water approaches a critical state when it gets close to 0 degrees Celsius. Cooling it from 1 degree Celsius to minus 1 degree Celsius has a much more drastic effect than cooling the same amount in the range of 20 degrees Celsius. In chapter 8, I will discuss a conjecture about the nature of the critical state that science and mathematics might evolve toward in the period that precedes a phase change. The concepts from information theory that are introduced and discussed in that chapter are essential to the development of this hypothesis.

A phase change has one characteristic property that is of central interest here: When a system is subject to a phase change, very little happens for a relatively long time. Then a very large change occurs very quickly, essentially instantaneously. This property has major implications for forecasting the future of a system. In particular, it means that any attempt to extrapolate the behavior of a system across a phase change is doomed to failure. In other words, if we use our experience with the behavior of a system before a phase change to try to understand how that same system will behave after the phase change, our expectations will not merely be wrong, they will not even be close. Although it may often be possible to extrapolate the behavior of the system through a period that falls between phase changes, it is never possible to extrapolate across any interval that spans a phase change.

It might seem that, rather than trying to use extrapolation techniques to deal with phase changes, it would be better to try to understand the underlying dynamics of the nonlinear system and thereby model and predict the properties of its phase changes. This appears to be a reasonable suggestion but in practice it does not work. Even the very simplest systems such as the freezing of water to ice are far too complicated to work out the details of the nonlinear theory. As Gunton et al. stated (1973, p. 270): "In spite of extensive experimental and theoretical studies of these first-order [phase] transitions, a first principles understanding does not yet exist." And if a theoretical understanding is lacking in the simplest of physical systems, then any attempt to work out a detailed theory of the much more complicated functioning of systems of human behavior is clearly hopeless. Still, it can be very useful to know that such complex systems are expected to exhibit phase changes or at least exhibit behavior that can be modeled accurately as a sequence of phase changes. Even though we cannot predict exactly what will happen across a phase change, we need to be aware that they can and will occur. The situation is perhaps analogous to exploring an unknown waterway in a canoe. It is useful to know that there are waterfalls ahead even when you do not yet know exactly where they will occur, or how high they are, or what lies beyond them.

It might be useful to describe a couple of instances of familiar historical events that can be thought of as examples of phase changes in the past. One such example involves the famous battle between the *Monitor* and the *Merrimack* in the U.S. civil war, the ironclad warships that permanently changed naval warfare. We can think of this as a phase change because no reasonable extrapolation of the capabilities of wooden ships in the past would give any clue to the capabilities of the *Monitor.* McPherson (1988, p. 377) quotes The London *Times* as commenting: "Whereas we had available for immediate purposes one hundred and forty-nine first-class warships, we now have two. . . . There is not now a ship in the English navy apart from these two that it would not be madness to trust to an engagement with that little *Monitor.*" (The Royal Navy had two experimental ironclads.)

Another example of a phenomenon that can be reasonably modeled as a phase change is the development of nuclear weapons in the middle of the twentieth century. Chemical explosives have been in use in Western civilization since about the fourteenth century when gunpowder was introduced. And the increase in explosive power for modern chemical explosives such as dynamite and TNT over old-fashioned gunpowder is less than one order of magnitude; furthermore, the change was spread out over a half-dozen centuries. But then overnight the introduction of atomic weapons produced an increase of some seven orders of magnitude, factors of tens of millions. It is hard to conceive of a more dramatic example of the fundamental property of a phase change: No reasonable extrapolation of the previous performance of chemical explosives would be anywhere close to the magnitude of atomic explosions.

In the discussion that follows I would like to turn away from chemical and atomic bombs and discuss instead something that is really explosive: information. Information is the root cause of many of the phase changes that are the main focus of this book. I am particularly interested in the effects of the information explosion that has been generated by the recent revolution in computer technology. As we shall see, this revolution is causing phase changes in almost every area of science and is reaching even into mathematics. And since the principal effect of a phase change is to make it very difficult to project the future course of events in a field, how can we proceed to try to understand the full impact of the computer revolution? One way would be by examining the effects of the phase changes in science and mathematics in the past. Contrasting the effects of these earlier phase changes to the effects of computer technology today will give us insights into the magnitudes of the effects of present-day phase changes.

For example, as I shall discuss in detail in chapter 2, the invention of the telescope caused a phase change in astronomy. The effects of this invention fit the critical definition of a phase change: No reasonable projection of astronomy from the pretelescope era comes even close to approximating astronomy in the posttelescope era. Similarly, in chapter 3, I will examine the phase change in biology caused by the invention of the microscope.

The phase changes in science and mathematics that will be the focus of the following chapters have one singular and very interesting feature in common: They all involve a technological or conceptual invention that gave us a novel ability to *see* things that could not be seen prior to the phase change. Here I am intentionally using the verb "see" in several different and important senses of that word. In the case of the invention of the telescope and the microscope, I mean the word "see" in its most literal sense. Both instruments expanded the capability of the human eye to see things that could not be seen without them. In the case of the use of X rays for medical purposes, the X-ray image is recorded on photographic film, yet this can also be regarded as an extension of our ability to see.

In the case of Ernest Rutherford's famous use of alpha particle radiation to probe the internal structure of the atom and discover the atomic nucleus, the word "see" is carried to another level of abstraction. It is reasonable to think of "seeing" the atomic nucleus using alpha particle radiation instead of light. Today Rutherford's insight has been carried much farther with the development of cyclotrons, synchrotrons, and other particle accelerators that modern physicists use to probe and "see" ever smaller components of the physical universe. And in a recent famous experiment, Luis Alvarez used cosmic-ray muons to "see" (or "x-ray") the internal structure of one of the Egyptian pyramids at Giza.

Finally, I mean to use the word "see" in its most general sense, to perceive and understand, as in the usage: "Aha! Now I see." In this sense of the word the development of Newtonian mechanics enabled us to "see" why Keplerian orbits are elliptical in shape rather than circular. If the earlier uses of the word "see" encompass the observational and experimental branches of science, then this one covers the theoretical side of science.

A novel capability to *see* things is often intimately connected to a similar novel capability to *do* things. Thus, the ability to put robot probes in orbit around the Moon was essential to developing the ability to see the lunar farside. And the cause-and-effect relations between the ability to do something novel and the ability to see something novel can be complex: Sometimes we can see things because we can do things, while other times we can do things because we can see things. The ability to orbit space probes preceded our ability to see the lunar farside, but Rutherford's ability to "see" atomic nuclei preceded his ability to split them. It is useful to recognize the intimate relation between a novel ability to see things and a novel ability to do things, but this discussion will focus largely on the first aspect, the ability to see things, because this capability seems more important to the development of science and mathematics. Still, we will not ignore the development of novel abilities to do things when they are relevant to developing new ways of seeing things.

Thus there are two important components to the definition of a phase change in science and mathematics: First, that any extrapolation of the previous behavior in the field will completely fail to give an accurate picture of the field following the phase change; second, the phase change is characterized by a novel ability to see things that could not be seen prior to the phase change. In the chapters that follow I will explore a series of examples of phase changes defined this way.

It is natural to wonder how the concept of a phase change relates to Thomas Kuhn's famous concept of a "paradigm shift" in science (Kuhn, 1970). The idea of a paradigm shift is important, although Kuhn does not define the word "paradigm" very precisely. As Weinberg noted (2001, p. 190): "Margaret Masterman [1970] pointed out that Kuhn had used the word 'paradigm' in over twenty different ways." Kuhn (1970, p. 10) comes closest to defining what he means by a paradigm when he describes it as something that "was sufficiently unprecedented to attract an enduring group of adherents away from competing modes of scientific activity. Simultaneously, it was sufficiently open-ended to leave all sorts of problems for the redefined group of practitioners to resolve." This definition could cover everything from Newtonian mechanics to polywater and cold fusion. For our purposes we can think of a paradigm as a set of fundamental ideas and theories that have broad acceptance. A good example of a paradigm shift might be the replacement of Newtonian mechanics by Einsteinian relativistic mechanics in the early decades of the twentieth century.

It seems to me that in developing the concept of a paradigm Kuhn has perhaps focused too closely on the theoretical side of science. This is natural for an individual such as Kuhn who was trained in theoretical physics. Indeed, the major difference between Kuhn's views and those expressed here may simply reflect the fact that much of my own career was spent developing observational techniques in radio interferometry rather than in theoretical work.

Perhaps Kuhn's most surprising misunderstanding of the nature of scientific research lies in a serious misapprehension of the basic motivation of most researchers. Kuhn dismisses "the excitement of exploring new territory" as something that "the individual . . . is almost never doing" (1970, p. 37). Instead, he says, "What then challenges him is the conviction that, if he is skillful enough, he will succeed in solving a puzzle that no one before has solved or solved so well" (Kuhn, 1970, p. 38). He then goes on to compare this puzzle solving to working on a jigsaw puzzle. Indeed, Kuhn devotes an entire chapter to "Normal Science as Puzzle-solving." (For a more detailed critique of some of Kuhn's ideas see Weinberg, 2001, pp. 187–206.)

Of course it is true that most researchers are good at puzzle solving, and most of them enjoy doing it. But too great a focus on merely solving puzzles leaves a researcher open to the charge of being "clever" in an almost pejorative sense of that word, a sense that suggests shallowness and a lack of understanding of deeper issues. Thus, although cleverness in solving puzzles is part of the intellectual armament of most researchers, it is usually far from being their fundamental motivation. That fundamental motivation is almost always a desire to see, to perceive, to understand things that had not been seen before. This compelling curiosity is the real force that drives phase changes in science and mathematics.

If we therefore seek to redress some of the imbalance in Kuhn's models by shifting our focus away from mere cleverness at solving puzzles and toward the fundamental curiosity that motivates most researchers, we see a rather different picture of the development of science than the one presented by Kuhn. This new picture focuses on phase changes, on radical improvements in our ability to see things, rather than on paradigm shifts. These phase changes are both less common and in some sense more fundamental than Kuhn's paradigm shifts. Indeed, as we shall see, phase changes in science tend to generate paradigm shifts. The reason is not difficult to understand: Following a phase change researchers exploit their new ability to see and discover things, and many of their new discoveries do not fit well into earlier paradigms. The paradigm shifts that follow are often simple and forced responses to the new observations and discoveries that follow a phase change. This does not mean that paradigm shifts are unimportant. Indeed, there would not be much point to a phase change without the succeeding paradigm shifts. But a phase change is a simpler concept than a paradigm shift, and it is one that is easier to define and identify in the historical record.

Indeed, the distinction between phase changes and paradigm shifts is closely related to the classical distinction between observation and experiment on the one hand and theory and interpretation on the other. But one of the important features of phase changes that we will explore here is that, because phase changes are related to improvements in observing techniques, they can often be quantified and intercompared quantitatively more easily than paradigm shifts, which are inherently more difficult to quantify. The fact that phase changes can often be quantified is one of the properties that makes them intrinsically interesting for study.

There is one other critical difference between a phase change and a paradigm shift: A phase change is often quite simple—you have a telescope or you don't; you have a microscope or you don't. In contrast, the development of a paradigm shift from a phase change can be quite complicated, involving a complex interplay of observation and theorizing, followed by more observation and theorizing. These complications are often case-specific and not easy to make generalizations about. As Weinberg noted (2001, p. 89):

> The interaction between theory and experiment is complicated. It is not that theories come first and then experimentalists confirm them, or that experimentalists make discoveries that are then explained by theorists. Theory and experiment often go on at the same time, strongly influencing each other.

The next chapters will explore some of the historical phase changes that have occurred in various branches of science and then compare and contrast them to the phase changes that are being generated today by the computer revolution. Computer technology will be seen to be the most powerful tool ever developed for expanding our ability to see things that were previously impossible to see. And it will do this in virtually every field of science and mathematics. This technology is therefore generating phase changes of a magnitude and at a rate that is utterly without precedent throughout the entire history of science and mathematics.

At the risk of seriously oversimplifying the discussion in the remaining chapters of the book, the main points there can be summarized as follows:

- The computer is more important to astronomy than the telescope.
- It is more important to biology than the microscope.
- It is more important to high-energy physics than the particle accelerator.
- It is more important to mathematics than Newton's invention of calculus.
- It is more important to geophysics than the seismograph.
- And so forth.

To make clear what I mean by these assertions I need to carefully define exactly what I mean by the word "important." An invention will be called important when it generates a phase change in science, and "more important" when it generates a larger phase change than some other invention to which it is being compared. I am not arguing that a phase change is more important than a paradigm shift. They are two different things that are both intrinsically important; to argue that one is more important than the other would be tantamount to arguing that an apple is more important than an orange. Similarly, because paradigm shifts are not easy to quantify, it is difficult to argue that one paradigm shift is more or less important than another. But phase changes are fundamentally different in this regard because phase changes can often be quantified. This makes it straightforward to argue that one phase change is more important (larger) than another. For example, the computer is more important than the telescope in astronomy because, as we shall see, the phase change caused by the introduction of computerized astronomical instrumentation is much larger than the phase change caused by the original invention of the telescope. In fact, computerized instrumentation sometimes outperforms conventional telescope instrumentation by as much as six orders of magnitude.

I am not arguing that inventions such as the telescope and the microscope were unimportant. On the contrary, they were among the most important inventions ever made prior to the computer. Neither am I arguing that they will not continue to be used in the future. Rather, I am simply arguing that the effect of computer technology will be very similar to the effect of any or all of these inventions in the past, but it will be vastly larger in magnitude. In astronomy, for example, the step from naked-eye observation to naked-eye plus telescope was smaller by many orders of magnitude than the corresponding step to naked-eye plus telescope plus computerized instrumentation. In chapter 2 I will explore a variety of different ways to make this statement quantitative. I will be able to put some explicit numbers on the effects of various types of instrumentation and analyze those effects quantitatively. Then in succeeding chapters I will explore ways to do the same type of analysis in fields other than astronomy.

With the beginning of the computer era we have come to the end of what we might call the "birch-bark canoe" era of exploration in science and mathematics. By this I mean the end of the era in which everything was done with hand labor, when instruments were operated by hand, data were recorded by hand, and calculations and analysis were done by hand. If we think of science and mathematics conceptually as a vast unknown continent that we are trying to explore, then the precomputer era is analogous to the period of exploration of North America by the great voyageurs who traversed the Great Lakes and the Mississippi valley and pushed into the Rocky Mountains entirely with hand labor (and perhaps a bit of horse labor). In sharp contrast, today's intrepid researchers are exploring new unknown continents with equipment analogous to jet aircraft with photo-reconnaissance apparatus. And this analogy seriously understates the case because the capabilities of jet aircraft are only about two orders of magnitude greater in speed and size than a birch-bark canoe. In contrast, the advantage of computerized technology over precomputer technology will generally entail increases of many more than two orders of magnitude.

I do not mean to denigrate in any way the great explorers of the birch-bark canoe era. On the contrary, I am awestruck by the superhuman feats accomplished by both the voyageurs in their real birch-bark canoes and by the great scientists of the precomputer era, people such as Newton, Euler, Gauss, Maxwell, Darwin, Rutherford, Hubble, and Einstein. These people accomplished incredible feats of insight, exploration, observation, and computation while working under the crippling handicap of having to do all their exploring, observing, and calculation by hand, without the benefit of computers. (Insight, however, is still done exclusively "by hand," without mechanical aid.)

With the dawning of the computer era and the end of the birch-bark canoe era, we are poised on the brink of the greatest revolutions ever seen in the history of science and mathematics. The excitement of these revolutions is captured in Wordsworth's verse quoted earlier (although Wordsworth was actually referring to quite a different revolution) (Wordsworth, 1979, p. 297). It is no great stretch to think that future generations will regard the computer revolution as the very beginning of both science and mathematics or, certainly, very much closer to the beginning than to the end. The following chapters will explore the excitement of some of these ongoing revolutions.

2

PHASE CHANGES IN ASTRONOMY

Astronomy may be the best field in which to begin the discussion of phase changes because the phase changes that occur in astronomy are so very dramatic. In addition, there is broad general interest in astronomy, so its phase changes are perhaps more familiar than the ones we will find in more esoteric areas of physics, mathematics, and even biology.

There are a number of phase changes in astronomy that I shall examine here that were spawned by inventions such as the telescope, the spectrograph, and photography. All of these inventions caused revolutions that match both parts of the definition of a phase change: First, that the change is so radical that no reasonable extrapolation of anything that happened before can come close to giving an accurate picture of what happens after the phase change; second, the phase change involves an ability to see things that could not be seen previously. And each of these inventions generated paradigm shifts.

Galileo is generally credited with being the first person to use a telescope for astronomical observations in the year 1609. With his new telescope he discovered a number of astronomical objects that had never been seen before, indeed, including many objects that could not be seen without a telescope. Perhaps most famously, he discovered four moons orbiting the planet Jupiter. He also found sunspots on the surface of the Sun and used them to make the first observations of the rotation of the Sun. He was the first to observe the phases of the planet Venus. In addition, he trained his telescope on the Milky Way and discovered that the familiar, faint, enigmatic glow that spans the sky is actually composed of countless stars that cannot be seen individually without a telescope.

Galileo's discoveries of the satellites of Jupiter and of the rotation of the Sun were relevant to one of the major and controversial paradigm shifts of the day, the shift from an Earth-centered Ptolemaic model of the universe to the Sun-centered Copernican model. Galileo argued that the Sun's observed rotation and the revolution of the Galilean satellites around Jupiter both gave indirect support to the Copernican viewpoint. His discovery of the phases of Venus was particularly damaging to the conventional Ptolemaic model, in which the Sun was not positioned correctly to account for these phases. Galileo also noted that the motion of sunspots could be used to define the solar equator, and the angle between this equator and the Earth changes in a manner that would be expected from the Copernican motion of the Earth about the Sun.

Following Galileo, astronomers began looking for direct evidence of the Copernican motion of the Earth about the Sun. The obvious phenomenon to look for was parallax, the apparent motion of distant objects caused by an actual motion of the observer. Parallax in planetary observations was an essential component of the Copernican model, which achieved its great simplicity compared to Ptolemaic models by replacing many of the Ptolemaic epicycles by the simple parallax effects of the orbital motion of the Earth.

But the Earth's motion should cause parallactic motions of stars as well as planets. These stellar parallaxes should have a one-year period with a well-defined phase if they are caused by the orbital motion of the Earth. Only the amplitude of the motion was unknown because the distances to the stars were unknown. Indeed, what astronomers of that time could not have known is that stars are so very distant that the amplitude of their parallactic motion is extremely small, typically less than one arcsecond and very nearly beyond the reach of telescopic observations. In fact, stellar parallax was not convincingly measured until Bessel's work in the nineteenth century. But in looking for parallax, astronomers accidently stumbled across another effect of the Copernican motion of the Earth, an effect called stellar aberration that happens to be independent of the distance to the stars.

Aberration was discovered by James Bradley in about the year 1725. While looking for stellar parallax effects, he found an annual motion of the star Gamma Draconis that had a magnitude of about

40 arcseconds. But he realized immediately that there was something seriously amiss with the motion that he measured for Gamma Draconis. The period was correct but the phase (direction) was wrong by about three months for the parallactic motions that should be produced by the known orbital motion of the Earth. It took Bradley several years to figure out that the motion he had observed was not an effect of the change in the Earth's position (parallax) but was rather an effect of the change in the Earth's velocity (aberration).

The aberration effect is easy to understand, although no one prior to Bradley had anticipated it. A simple analogy can be found in the apparent motion of raindrops. If raindrops are falling straight down, but you observe them from a vehicle moving horizontally at about the same speed as the raindrops, they will appear to you to be moving not vertically but at an angle of about 45 degrees to the vertical, because you observe the sum of the velocity of the raindrops plus the velocity of the vehicle. Similarly, the orbital velocity of the Earth changes the apparent direction of incoming light waves. Unlike the tiny parallax angles, the magnitude of aberration is tens of arcseconds, well within the capabilities of telescopes, although still invisible to the naked eye. Once again the telescope allowed us to see something (stellar aberration) that could not be seen without it, and the new observations gave the first direct evidence of the Copernican orbital motion of the Earth.

One of Galileo's other observations, that the Milky Way consists of countless numbers of faint stars, led directly to another paradigm shift in astronomy. If the Copernican paradigm shift concerned the position of the Earth relative to the Sun, the new paradigm shift concerned the position of the solar system in the universe.

If you look at the stars that can be seen with the naked eye you will find them approximately evenly distributed about the sky, which is roughly what you would expect in a Copernican universe with the solar system at its center. But Galileo's observation showed that stars are not uniformly distributed across the sky, but instead are concentrated in a fairly flat disk that we call the Milky Way. The question of the distribution of stars in the universe soon became a matter of intense astronomical as well as philosophical debate.

In the eighteenth century William Herschel made the first systematic attempt to measure the distribution of stars in the Milky Way. He began a detailed count of stars that could be seen with a telescope in various directions around the Milky Way's disk. Herschel reasoned that if the Sun were located off-center in a disk of stars then you would count fewer stars when you looked in the direction toward the closest edge of the disk. Herschel observed that the numbers of stars were roughly the same in every direction around the Milky Way, and he concluded that the Sun was located near the center of the Milky Way. This was a good first try at the problem of the location of the solar system, but it came up with an answer that was wrong. The Sun is not at all close to the center of the Milky Way. What Herschel failed to realize is that dust contained in the disk of the Milky Way blocks the light from distant stars, so Herschel's star counts did not extend to stars that were even close to the edge of the Milky Way in any direction.

The actual position of the solar system in the Milky Way was not resolved until Shapley's telescopic observations of the distribution of globular clusters in the early twentieth century. These globular clusters are enormous clumps of stars that are distributed in a spherical "halo" around the Milky Way's disk. Because they are outside the plane of the Milky Way, they are relatively free of obscuring dust and can be observed clear across the galaxy.

This paradigm shift concerning the position of the Sun in the universe continued then with the obvious question of whether the Milky Way was the only disk of stars in the universe. In particular, astronomers knew of some peculiar objects that they called "spiral nebulae," whose structures had been observed visually with Lord Rosse's 72-inch diameter telescope in the nineteenth century. These spiral nebulae looked suspiciously similar to what we might expect a distant "Milky Way" to look like.

In the twentieth century enormous telescopes were constructed explicitly for the purpose of investigating the question of the nature of the spiral nebulae. These new telescopes included the famous 100-inch telescope at Mt. Wilson and the 200-inch telescope at Mt. Palomar. The puzzle was resolved when Edwin Hubble used the Mt. Wilson telescope to "see" individual stars, including the famous Cepheid variable stars in the Great Spiral Nebula in Andromeda. Hubble had discovered that the Milky Way is not alone. The universe is speckled with countless billions of other disk-shaped and elliptical-shaped and even irregular groupings of stars that were given the name "galaxies." This switch from the Copernican idea of the Sun at the center of the universe to an understanding of the Sun's position in the Milky Way and the Milky Way's position as one among billions of galaxies was one of the major paradigm shifts in the history of astronomy. For a discussion of a similar and closely related paradigm shift related to the size of the universe (rather than to the Earth's location), see the discussion in Robertson (1998, pp. 38–39).

There are obviously many additional examples of things that were discovered using a telescope that could not have been seen without it. Herschel's discovery of a new planet, Uranus, was followed by the discovery of Neptune and Pluto. In 1802, the first asteroid was discovered, and today over 20,000 asteroids are known. Also, many more moons were discovered in orbit around Mars, Jupiter, Uranus, Neptune, and even Pluto. And the discovery of the nature of spiral nebulae has already been recounted. Following the invention of the telescope, astronomers discovered that the universe is full of novel objects that cannot be seen with the naked eye. Important paradigm shifts resulted from this phase change that followed the invention of the telescope. Why then should we think that the computer revolution is going to cause an even greater phase change in astronomy than the invention of the telescope? To answer this question we need to examine the effects of the invention of the telescope in a little more detail.

The telescope had two major effects on astronomical observations: First and most obvious, it allowed astronomers to see celestial objects with higher angular resolution (greater magnification). Second, it improved light-gathering power so that astronomers could see faint objects. In other words, astronomers using a telescope could see objects that are too small and/or too faint to be seen with the naked eye. The fundamental point of the argument here is that computer technology is producing even larger increases in both angular resolution and in light-gathering power than the original invention of the telescope did.

Angular resolution in the pretelescope era was limited to about one minute of arc, the smallest angle that can be resolved by the unaided human eye. The telescope improved this to about one second of arc, the limiting resolution that is possible for a telescope that is operating within the variable refractivity of the Earth's atmosphere. This variable atmospheric refractivity causes the familiar "twinkling" of stars, and it blurs their positions and features at about the level of one arcsecond. Thus, the telescope provided an improvement in magnifying power of about a factor of sixty. But computer technologies have allowed the development of a variety of techniques that produce much greater increases in magnification. For example, adaptive optical techniques are able to use computers to control arrays of piezoelectric crystals that bend or distort an optical surface in such a way as to correct for atmospheric refraction fluctuations in real time. Adaptive optical techniques are able to improve angular resolutions to about a hundredth of a second of arc, a factor of a hundred or so beyond what can be done with uncorrected telescopes. Another way to defeat atmospheric refractivity problems is to use computerized telescopes in orbit (such as the well-known Hubble Space Telescope) that also achieve resolutions of a few hundredths of a second of arc. Another strategy involves shifting to a different observing frequency where refractivity is more tractable: Computerized radio interferometry, for example, routinely achieves an angular resolution better than a thousandth of a second of arc. And computerized optical interferometers in space (that are presently on the drawing boards) should achieve a resolution of about a ten-thousandth of a second of arc. To put this in perspective, a ten-thousandth of a second of arc is about the width of a pencil in Honolulu as seen from New York.

Thus, the invention of the telescope improved angular resolution by a factor of about 60, but computerized observing techniques are pushing an additional factor of a hundred, a thousand, ten thousand. And the full effect of these increases in resolution should be calculated as the square of these numbers because the effective "area" that can be seen in the sky is proportional to the square of the angular resolution. This suggests that computer technology is capable of producing increases in the number of previously invisible objects that we might now expect to see by factors of 10^8 or so, a factor of about a hundred million compared to the factor of 60 squared (3600) for the original invention of the telescope.

The improvement in light-gathering power provided by computers is less dramatic than the improvement in angular resolution, but it is still very significant, involving factors of 10 or so. Light-gathering power is directly proportional to the area of the telescope's primary lens or mirror (or, equivalently, proportional to the square of the primary's diameter). And the process of increasing the diameter of primary mirrors had run into a major stumbling block in the precomputer era.

In 1948, in one of the greatest triumphs of precomputer astronomy, the great 200-inch Hale telescope was dedicated on the summit of Mt. Palomar near San Diego. For decades the Mt. Palomar telescope was the largest optical instrument on Earth, not because astronomers did not want to build larger telescopes, but because 200 inches is close to the maximum size that is possible for a telescope that is built and operated without computer technology. Actually, the 200-inch telescope may be a bit beyond those limits. The 200-inch telescope was never quite able to achieve the level of optical performance that its designers had hoped for.

The key problem in building a very large telescope is simply this: In order to focus a sharp image, the optical surfaces must remain rigid to within a fraction of a wavelength of light, about 60 millionths of a centimeter for yellow light. The Mt. Palomar reflector (and all the telescopes that preceded it) had achieved the necessary rigidity by using the sheer physical strength of glass.

But at 200 inches diameter (about 5 meters or 17 feet), the Mt. Palomar mirror is about as large as you can make a glass mirror and still have the necessary rigidity without using computer technology. A larger mirror will bend under gravitational, thermal, and wind stress; its intrinsic strength is not sufficient to maintain the necessary level of optical tolerance. And the various fixes to this problem involving larger and more massive or more clever supporting mechanisms quickly become very costly. It is therefore impractical to build a glass mirror much larger than 200 inches in diameter that will meet the exacting standards of astronomical optics without using computer technology.

But today computer technology is being exploited to successfully construct and operate telescope mirrors that are two or three times larger than the Mt. Palomar reflector at cost levels that are manageable. These new mirrors are supported on arrays of electronically actuated pistons that are controlled by a computer. The computer calculates the gravitational and wind stresses that arise with a particular tilt of the mirror and adjusts the supports to maintain the mirror in its optimum shape. Some particularly thin glass mirrors are being designed today that are so far from being physically rigid that they are referred to as "floppy" mirrors. Other reflectors, such as the brilliantly successful Keck telescope in Hawaii, use more than one piece of glass for the main mirror. The main reflector of the Keck telescope consists of 36 separate mirrors, each one held and constantly adjusted in position under computer control to maintain an optimum image. With this technology, telescopes with 10-meter-diameter mirrors have been constructed and operated, 25-meter telescopes are on the drawing boards, and there are serious discussions of a telescope in the 100-meter-diameter range, the OWL telescope ("OWL" stands for "OverWhelmingly Large").

Radio astronomers also found themselves facing very similar problems with the construction of large radio telescopes. Current designs for parabolic radio antennas ("dishes") in the 100-meter-diameter range have pushed the limits of both the physical strength of metal structures and the budget strictures of research organizations. The new 100-meter radio telescope at the National Radio Astronomy Observatory in Greenbank, West Virginia, makes extensive use of computerized control of the shape of the dish. But radio astronomers are taking a different tack as well. As Robert Irion (2002) reported about the Allen Telescope Array (ATA), an array of radio telescopes recently constructed in northern California:

> The array's concept—many receivers linked by cheap electronics—is rumbling through radio astronomy. By cutting the structural costs of huge dishes and instead combining signals from many small detectors, radio astronomers aim to explore the cosmos electronically at a non-astronomical price.
>
> "In a metaphorical sense, we're learning how to build telescopes out of computers, not metal," says physicist Kent Cullers of the SETI [Search for Extraterrestrial Intelligence] Institute in Mountain View, California. "This is the future of radio astronomy," agrees astronomer Leo Blitz of the University of California (UC), Berkeley, which is building ATA with the SETI Institute. "If you want to build a large telescope for a fixed amount of money, I can no longer think of any reason to build a single large dish."
>
> Cullers and Blitz are confident that 12 to 15 years hence, this trend will produce a telescope of breathtaking scale: the Square Kilometer Array [SKA]. As its name implies, this instrument would gather waves with a combined detecting area of a full square kilometer—making it 100 times more sensitive than any existing array. SKA would expose the now-invisible era when hydrogen first clumped together [in the primordial universe], tracing the "cosmic web" of dark matter that underlies all structure in the universe. Other studies unique to SKA's radio window include mapping magnetic fields in and among galaxies in exquisite detail, finding thousands of pulsars and using them to track gravitational ripples in space, and extending the search for intelligent life to tens of millions of stars.

Calculating the magnitude of the phase change in astronomy that has been caused by computer technology in terms of the increase in angular resolution and/or light-gathering and radio-energy gathering power, as we have done here, presupposes that there will be something new to see at higher powers. If there is nothing but blank sky beyond some limiting angle or brightness, then increasing the resolution (magnification) or the sensitivity will not show anything new. And, in fact, there is no way to know in advance that increasing resolution or sensitivity will actually allow us to see anything new. But the universe has never let us down in the past. Increasing resolution has always presented new and remarkable objects for study. There is no reason to expect that this will change in the future, although there is no way to be certain. In terms popularized by modern chaos theory, the universe appears to be scale-invariant. In other words, a wealth of new phenomena is available to be observed at every scale of distance and angle and brightness. One recent experiment, the Hubble Deep Field image, used 10 days of precious observing time on the orbiting Hubble telescope to look at a blank part of the sky, to test whether the higher angular resolution and sensitivity of the Hubble instrument might allow us to see things in what appeared to be blank sky with conventional instruments. The area of blank sky was devoid of the obscuring effects of nearby objects (or it would not appear blank in ordinary telescopes) and might allow observations clear across the entire universe. The experiment succeeded beyond all expectations. It not only yielded a wealth of information about some of the most distant galaxies ever observed but also fortuitously caught the light from the most distant supernova ever seen. This supernova observation gave extraordinary new information about the expansion of the universe, which now appears to be increasing with increasing distance.

Thus, our experience tells us that we should expect that by increasing our power to see things we will continue to find a wealth of new phenomena to observe. And similar assumptions about scale-invariance of the existence of interesting phenomena will also be made for the other phase changes discussed in this chapter, as well as those in biology and physics and the other topics of the succeeding chapters.

The introduction of photographic techniques in the nineteenth century produced another phase change in astronomy. In addition to providing a physical record of an astronomical observation (a nontrivial advantage for this technology), photographic plates gave astronomers another major increase in light-gathering power. Instead of simply increasing the diameter of the telescope to collect more photons all at once, astronomers could let a photographic plate collect photons over time intervals of hours and again develop an ability to see objects that were too faint to be observed with the naked eye at the telescope eyepiece. The ability to observe faint objects using long photographic exposures was critical to Hubble's work on observing stars in the Spiral Nebula in Andromeda that resolved the question of the nature of external galaxies.

But prior to the computer revolution, astronomers were severely limited by the properties of photographic plates. Photographic plates, for all their remarkable chemistry, are simply not an efficient medium for recording incident light. This lack of efficiency is the principal reason that conventional photographers are often forced to use flashbulbs and other bright light sources. Photographic emulsions typically have an efficiency of 5% or less. In other words, standard photographic plates record only about one incoming photon for every 20 that strike the photographic surface.

Computer technology has spurred the development of efficient electronic detector chips to replace photographic plates. Special silicon chips called charge-coupled devices (CCD's) that were originally developed to be computer memory devices are able to detect 50 to 70% of the incoming photons. This is more than ten times the fraction that are recorded with a photographic plate. In effect, every telescope today that uses CCD technology has had its diameter enlarged by a factor of roughly 3 (the square root of 10). Thus, thanks to computer technology, the Mt. Palomar telescope is presently functioning as well as a 600-inch-diameter telescope would have in the precomputer era.

Photographic technology has other serious disadvantages in addition to its lack of sensitivity. The response of photographic emulsions to light is strongly nonlinear, and it suffers from what is termed "reciprocity failure." The word "nonlinear" means simply that a portion of a photographic negative that receives twice as many photons as another part will generally be more than twice as dark. The technical reason for this is that the chemical process of exposing a photographic emulsion is self-catalyzing. In other words, a plate that is partly exposed to light becomes more sensitive to additional incoming light. And the problem of reciprocity failure is a similar nonlinearity across time rather than across area: A photographic negative that receives 20 photons in two seconds will show a different exposure than one that receives the same 20 photons spread over two minutes of time.

In contrast, CCD chips and other modern detectors are quite linear, both spatially and across time. In other words, they can count photons more accurately than photographic technology does. With computerized detectors the precise measurement of the intensity of observed light is far easier and more accurate than was ever possible with conventional photographic technology. And because the CCD images originate in digital form they can be processed immediately by computers, sometimes even while the time exposure is still in progress. This allows astronomers to implement a number of novel observing techniques, one of which will be described below.

Yet another phase change in astronomy was caused by the introduction of the spectrograph. With a spectrograph an astronomer could see not only the location or direction of an incoming photon, but he could also see its wavelength (color). This was important for at least two reasons. First, individual atoms or molecules emit light at characteristic spectral lines or colors. Thus, a spectrograph allows the astronomer to identify or "see" the chemical properties of the objects being observed. This ability led immediately to a radical new paradigm concerning the abundance and distribution of chemical elements in the universe, thanks largely to the work of Cecilia Payne-Gaposhkin at Harvard in the 1920s. Prior to Payne-Gaposhkin's work, astronomers believed that the abundances of chemical elements that were observed on Earth were typical of the rest of the universe, and indeed, they had no reason to think otherwise. Payne-Gaposhkin showed that, in fact, most of the matter in the universe is hydrogen (73%), most of the remainder is helium (25%), and the more familiar oxygen, silicon, aluminum, iron, calcium, carbon, and other elements that are common on Earth compose only about 2% of the mass of the chemical elements in the universe.

Another paradigm shift spawned by the phase change that was touched off by the spectrograph came about because a spectrograph is able to measure the radial component of the velocity of an astronomical object relative to the velocity of the observer. The characteristic wavelengths of spectral lines of atoms and molecules are Doppler-shifted by an amount that is proportional to the radial velocity. By measuring Doppler shifts of spectral lines in distant galaxies, Edwin Hubble made what many consider to be the most important discovery in astronomy in his century. He found that these Doppler shifts were all in the same direction, toward the red end of the spectrum, and the amount of the shift was proportional to the distance of the galaxy. Today this Doppler shift is referred to as the cosmological redshift. In other words, galaxies are all moving away from us (except for some that are very nearby), and the magnitude of the redshift can be used to measure the distance to the object. Hubble had discovered that the universe as a whole is expanding.

Already spectroscopy is being combined with the new computerized telescopes in the 10-meter class and computerized detector technology to allow astronomers to tackle research projects that would have been difficult to impossible without these magnificent new instruments. The new projects include searches for planets around stars other than the Sun. Jayawardhana (2002, pp. 35–36) reports that

> Geoffrey Marcy (University of California, Berkeley) . . . leads the world's most prolific extra-solar planet research team. He and his colleagues break down starlight into its individual wavelengths and then look for lines in the spectrum that sway as the star wobbles due to an unknown planet's gravitational tug. . . . "Ultimately," Marcy declares, "it's all about photons. The more photons we collect—the more light we gather from our target stars—the easier it is for us to look for planets in their midst. It's truly awesome what Keck is doing for us. It's improved our precision tremendously."
>
> Within a few years Marcy hopes to find planets only a few times more massive than Earth. "That would be cool," he beams, "because those planets are likely to be rocky, where water could puddle into streams, lakes, and oceans, and water could really be in liquid form and act as the solvent for biochemistry." Finding wet, habitable worlds around other stars is an important step toward answering the grand question of extraterrestrial life.

Other important work using these new telescopes involve studies of bodies at the edge of our solar system. Again from Jayawardhana (2002, p. 37):

> Caltech's Michael E. Brown uses Keck to study Kuiper Belt objects (KBOs)—icy bodies within the outer edges of the solar system. KBOs range in diameter from just a few kilometers to 1,000 km or more. "They are the true primordial remnants of the early days of the solar nebula," says Brown, "They have been in a deep freeze for 4½ billion years." Brown uses Keck to obtain spectra that provide clues to their makeup. "These things are so faint that you just can't do it without a 10-meter telescope. Already Brown and others have found evidence of organic molecules in these primitive bodies.

But again, prior to the invention of computer technology, astronomers were limited to the capabilities of conventional spectrographs. The problems include more than just the limits of the photographic technology that was once used to record the astronomical spectra. Conventional spectrographs were also limited to recording the spectra of single small objects. If an astronomer tried to obtain the spectrum of the light from a large diffuse object such as the Spiral Nebula in Andromeda, the spatial extent of the object would blur the spectral lines, unless the observer masked off the light from all but a small portion of the object. Thus, spectroscopy was a long and tedious process, and one that consumed large quantities of valuable telescope time to get data from statistically large collections of objects. Today chipmakers are working on new classes of solid-state detectors that will record not only the location of photons but also their wavelength at the same time. Some otherwise serious and sober astronomers are positively salivating at the thought of being able to measure the spectra of objects across an entire field of observation. They have good reason to expect that they will find things that would otherwise have been overlooked; they will see things that have never been seen before.

This is still only a small part of the story of the impact of computers on astronomy, which ranges far beyond the simple ability to make telescopes that can detect smaller and fainter objects and analyze their spectra easier and faster. For one example, computerized astronomical databases are being constructed that promise to revolutionize astronomy in a completely novel way. One of the most dramatic examples is the Sloan Digital Sky Survey (SDSS), whose objective is to measure literally everything brighter than twenty-third magnitude in a swath that covers nearly one-quarter of the entire sky. ("Magnitude" is a technical term for brightness on a logarithmic scale. For the discussion here all we need to know is that larger magnitude numbers represent fainter objects. Sixth magnitude is close to the limiting brightness that can be perceived by the human eye, and twenty-third magnitude is about six million times fainter.) The SDSS is expected to measure the brightness of 100 million objects and cosmological redshifts for one million galaxies and 100,000 quasars. It should produce 40 terabytes (40×10^{12} bytes) of data over its five-year life. Already, with less than one-twentieth of its observing completed, it has discovered more than 13,000 quasars, including the most distant ones yet known, some with cosmological redshift values that are greater than six. Szalay commented about this and similar surveys: "Astronomy is about to undergo a major paradigm shift. . . . In a few years it may be much easier for astronomers to 'dial up' a part of the sky when they need a rapid observation, rather than wait for several months to access a (sometimes quite small) telescope" (1999, p. 54).

Beyond the Sloan Sky Survey even more ambitious projects are in the works. The Large Aperture Synoptic Survey Telescope (LSST) is a project that will construct an 8-meter-class optical telescope and use it to produce a complete image of the sky every week, recording five terabytes (5×10^{12} bytes) of data every night. The idea is to provide a complete record of transient phenomena from asteroids to supernovae that can be searched and studied. Back in the 1950s, the Palomar Observatory attempted a similarly ambitious project to photograph the entire night sky with conventional (precomputer) technology. The project took seven years and the resulting Palomar Observatory Sky Survey (POSS) was a vital reference work for astronomers for decades. The POSS was so successful that it was repeated with more sensitive photographic plates about three decades later, and the POSS II plates were converted to digital form, resulting in about three terabytes of digital data. Thus, the LSST is going to produce more data in a single night than the Palomar Sky Survey was able to do with seven years of effort.

Possibly the most ambitious projects along these lines are several efforts in the United States and Europe, whose objective is to provide on-line access to enormous astronomical databases. The projects are referred to as the National Virtual Observatory (NVO) in the United States and the Astrophysical Virtual Observatory in Europe. The idea is to make data archives easily accessible to researchers throughout the world. As one researcher put it (quoted in Feder, 2002, p. 20):

> "The virtual observatory concept is a very ambitious project and a great challenge to astrophysicists, mathematicians, statisticians, and specialists in all areas of computer science to work together," says Wolfgang Voges of the Max Planck Institute in Garching. . . . "There will be a fresh wind blowing through the graveyard of old and unused data. And there should be no newly formed graveyards in the future if we succeed in finding a globally accepted data format."

Computers also allow completely novel observing techniques that would have been inconceivable in the precomputer era. For example, some cosmologists posed the question of whether significant numbers of planet-size bodies might be found in the spherical "halo" that surrounds the Milky Way galaxy. With precomputer technology there was no way to answer this question. The problem, of course, is that planet-size bodies are rather difficult to detect at the distance of the nearby stars. How could they possibly be seen at distances that are tens of thousands of times greater? They would be too small and faint to be seen with the largest conceivable telescope that might be built. But it would still be possible to detect them if one of them happened to move directly between an observer and a distant star. When that happens the gravity field of the planet-size body would act as a lens that would amplify the light from the distant star. This "gravitational lensing" was predicted by Einstein.

This idea led to the MACHO project (MAssive, Compact Halo Objects). To detect these objects, all you have to do is watch a distant star and wait until a small body happens to come directly between you and the star. Of course, the odds are millions to one against this ever happening. The way to deal with such long odds is to use computerized telescopes and detectors to observe the brightness of millions of stars in a single night. You then repeat the same observations night after night after night. Without computers this would be an impossible task, but it is straightforward, almost easy, with automated telescopes and computerized CCD detectors. As a bonus, the study provided massive statistics on the properties and variability of the millions of ordinary stars that it monitored every night. And in 18 months, the MACHO project detected dozens of the planet-size bodies that it was looking for.

Similar automatic computerized searches are already being used to make direct observations of asteroids inside our own solar system. These searches are so effective that asteroid number 20,000 was discovered in the year 2000. What makes this accomplishment particularly impressive is that asteroid number 10,000 had been discovered in the previous year, 1999. Some of these searches are particularly focused on asteroids whose orbits cross the Earth's orbit. Such asteroids could impact the Earth and cause a cataclysm on the scale of the asteroid impact that apparently destroyed the dinosaurs. Other computerized searches have discovered dozens of bodies orbiting beyond Neptune, some of which are dubbed "plutinos" ("little Plutos"). Still other automated searches watch for supernovae in distant galaxies and are instrumental in determining the overall size and structure of the universe.

Some of the ways that astronomy has been affected by computer technology have been completely unexpected. Perhaps one of the most curious stories involves the discovery of an observation of Neptune by Galileo. The story begins with the work of Steven Albers, who was a student working at NASA's Jet Propulsion Laboratory (JPL) in Pasadena in the 1970s. Albers was interested in planetary occultations, a rare phenomenon in which one planet passes directly in front of another as seen from Earth. He used a then-powerful mainframe computer at JPL to calculate the time and position of every planetary occultation that occurred between the years 1557 and 2230, a total of 21 occultations. He was interested in the possibility that some of the historical occultations might have actually been observed. In 1979, he published his occultation list in *Sky and Telescope* to invite other astronomers to investigate this possibility.

Two researchers from Caltech, Stillman Drake and Charles Kowal, noticed that Albers's list included an occultation of Neptune by Jupiter in 1613. They realized that Galileo was making his celebrated observations of the satellites of Jupiter during this time, and furthermore, they knew that Galileo's original observing journals are preserved in the Central National Library in Florence. Galileo's handwritten journals contained sketches of Jupiter, its satellites, and occasionally some surrounding stars during this time. Drake and Kowal found that for the critical nights in 1612 and 1613, the journals contain sketches and careful measurements of the position of Neptune, which appeared to Galileo as a faint star. On one night Galileo's notes even comment that Neptune appeared to have changed its position relative to a nearby star since the previous night. Galileo apparently suspected that this object was not a fixed star, but with his primitive observing equipment he was unable to follow up on the discovery or even to find Neptune again after Jupiter had moved far enough away from Neptune to place it outside of the field of view of his telescope.

Drake and Kowal note that this observation occurred 234 years before the "official" discovery of Neptune in 1846, and they comment that they have had difficulty fitting Galileo's observed position of Neptune to an orbit calculated backward from modern observations (calculated by numerical integration on a computer, of course). Galileo's observation indicates a need for a slight adjustment to the orbital parameters of Neptune. (It would have been rather extraordinary had it not done so.) The record of this sighting of Neptune had languished unnoticed in the library in Florence for three and a half centuries. Without computer technology it might have remained unnoticed forever.

Other phase changes in astronomy were caused by the invention of radio, X ray, gamma-ray, and neutrino astronomy. Again, all these inventions enabled us to see things that could not be seen without them. And one of the potentially most important paradigm shifts in all history could involve a combination of computers with radio techniques in the famous SETI project, the Search for Extraterrestrial Intelligence. If this search were to be successful in detecting the presence of intelligent life elsewhere in the universe, it is generally acknowledged that it would be one of the most important discoveries of all time.

Radio observations are one of the obvious ways to search for extraterrestrials that happen to be technologically adept.

Already the Earth itself is surrounded by a halo of its own radio signals that is many tens of light-years across. And if other civilizations have had radio-transmitting technology for longer, perhaps half a million years, then their transmissions might span the entire Milky Way galaxy. But detecting these signals is very difficult in the absence of any a priori information about the direction and amplitude and frequency and modulation of the signals. An enormous amount of computer power is needed to search all possible frequency bands for artificial signals. Today the necessary level of computer power has been found at very low cost with the clever SETI@HOME project.[1] With this project, millions of individuals are donating time on their personal computers to help process the SETI data. They download the software and samples of the data and let their home computers process the data. The results are uploaded back to the SETI researchers. This technique of using otherwise wasted home computer power has proven so powerful that other researchers are emulating it to solve a variety of problems even in pure mathematics.

Additional uses of computers in astronomy involve numerical integration of spacecraft and planetary orbits and studies of the stability of the solar system. Such studies indicate that the behavior of the planets is chaotic on long timescales. Computerized numerical integration and nonlinear modeling is also used to study important effects such as supernova explosions and solar flares. Computers are also used to analyze data about Doppler shifts in sunlight caused by oscillations of the solar surface. These observations of the vibrations of the solar surface allow researchers to make detailed models of the physical state of the solar interior that controls or determines the vibration modes in a new research area called helioseismology. Still other uses involve operating robots that can send images back from the surface of Mars. The full uses of computer technology in astronomy have only begun to be probed.

This discussion has covered only a tiny sample of even the present uses of computers in astronomy. A full catalog would fill a number of volumes. The impact of the computer is perhaps analogous to the introduction of the telescope, the spectrograph, photography, Newtonian dynamics and gravity theory (see chapter 5), plus Brahe's observational data, all at the same time and on a larger scale. And this is a pattern that in succeeding chapters we will see repeated through other areas of science and even mathematics.

3

PHASE CHANGES IN BIOLOGY

After astronomy, biology is one of the best fields in which to explore the phenomenon of phase changes in the sciences because the invention of the microscope has close parallels to the invention of the telescope. Indeed, in many ways a microscope is merely a telescope that has been inverted: With a simple two-lens telescope the eyepiece is the smaller lens, but with a two-lens microscope the eyepiece is the larger lens.

The invention of the microscope produced a phase change in biology very similar to the phase change produced in astronomy by the invention of the telescope, and it did so for the same reason: The microscope enabled researchers to see things that could not be seen without it. And just like the telescope, the microscope spawned a sequence of paradigm shifts because many of the things that were seen for the first time simply did not fit well into earlier paradigms.

Before the microscope was invented biologists were limited to observing at the level of whole organisms and their major components such as bone, muscle, organs, and so forth. The entire ecosystem of single-cell organisms was completely unrecognized; indeed, its very existence was not even suspected. The invention of the microscope allowed biologists to see structures at the level of individual cells instead of the level of macroscopic organisms and components.

Galileo was involved in the early use of the microscope as well as the telescope, although he was probably not the original inventor of either instrument. He observed small insects using one of his famous telescopes modified for use in the reverse direction. But Galileo's microscopic observations seem to have received considerably less attention than his more famous telescopic observations of the heavens, perhaps because they did not provoke a charge of heresy.

In 1651 William Harvey published crude microscopic studies of a chicken's egg that challenged Aristotelian ideas about reproduction. Nearly two centuries were to pass before Von Baer discovered the mammalian ovum (egg) in 1827.

In 1665 Robert Hooke created a sensation when he published results of studies using a primitive microscope. Hooke's manuscript included hand drawings of small insects such as fleas. He also observed the cellular structure that he found in thin slices of cork.

Hooke's contemporary, the gifted Dutch researcher Anton van Leeuwenhoek, was perhaps the most important figure in the early history of the microscope. Leeuwenhoek developed a simple (single-lens) microscope that was far more powerful than Hooke's or Galileo's compound (two-lens) instruments. He used this new invention to make a series of enormously important discoveries, including the first observations of bacteria, red blood cells, spermatozoa, and the capillary blood vessels that connect arteries to veins. Leeuwenhoek's discovery of capillaries was of central importance to one of the major paradigm shifts initiated by Harvey, the idea of the circulation of blood, and his observations of bacteria led a century or so later to another paradigm shift in biology and medicine, the germ theory of disease developed by Pasteur and Koch.

In the 1850s Rudolph Virchow used the detailed microscopic observations of animal and plant tissues by Schwann and Schleiden to develop the cell theory. This theory proposed that every organism consists of an assemblage of one or more cells, each of which bears in itself the complete characteristics of life. The cell theory marked another of the major paradigm shifts that altered the very foundations of biology. It produced a framework that unified and clarified the basis not only of both plant and animal life but of single-cell life forms as well. Indeed, the cell theory clarified the basic underlying similarities that link every life-form that is more complicated than a virus.

The microscope further enabled researchers to see organelles inside eukaryotic (nucleated) cells, including mitochondria, chloroplasts, and the cell nuclei themselves. Leeuwenhoek had reported the first observation of the nucleus of a cell in 1702. Further study of the nucleus led to a phenomenological understanding of the biological processes of cell division, mitosis and meiosis. With these discoveries biologists were poised at the edge of resolving a set of fundamental problems whose importance had been recognized as far back as Aristotle, a detailed understanding of the reproduction and growth of biological organisms.

Thus the phase change in biology that is associated with the invention of the microscope was central to at least three major paradigm shifts: the circulation of blood, the germ theory of disease, and the cell theory. It also produced the beginning of a fourth paradigm shift, the first understanding of the reproduction and growth of life-forms. But the full development of this fourth paradigm shift would require another phase change in biology, a change from seeing things at the level of cellular structures to seeing things at the level of atomic and molecular phenomena. Light microscopes have an unavoidable and fundamental limitation that prevents them from being used to observe structures at the atomic or molecular level. The problem is that the wavelength of visible light is several thousand times larger than the typical sizes of atoms. Visible light is therefore too coarse an implement to allow us to "see" or resolve individual atoms and molecules.

The next phase change in biology, the shift to observing molecular structures, involved the development of a range of new technologies in which computers have a central role. Not only do many of the new instruments require computer technology in order to operate at the molecular level, but also the quantity of information that is produced by these same instruments is so vast that it would be largely useless if we did not have the raw calculating power of computers available to analyze it.

The magnitude of the earlier phase change that was touched off by the invention of the microscope involved an increase in magnification of about a factor of 500 to 1,000 beyond the limits of what can be seen with the naked eye. The magnitude of the new phase change from the cellular level to the atomic and molecular level involves an additional magnification factor of about 5,000, about one order of magnitude larger than the phase change associated with the invention of the microscope. But just as with magnification (angular resolution) in astronomy, the real size of the phase change is the square of these numbers because microscopes are generally used to magnify areas. Thus, this new phase change is about two orders of magnitude larger than the one associated with the invention of the microscope.

As noted above, the phase change involved in the shift to the molecular level in biology required the development of several new technologies. The story is complex and computer technology is of central importance. For example, not only would the human genome project have been impossible without computer power, but also, as we shall see, the information coded into the genome would be nearly valueless without the extraordinary power of computers to search out and analyze the functional components of the genome. And recent proposals to analyze the structure and folding of proteins may become the largest single user of computer power in history. George Bell describes the need for computational power in molecular biology (1988, p. xiii):

> The fields of molecular biology and genetics are faced with the accumulation of quantitative information at an ever increasing rate such that the unaided human mind cannot begin to assimilate or analyze its significance. The chief source of this information is DNA sequencing and, thereby, the associated sequences of amino acids in proteins; but genetics, macromolecular structure and other data sets are also large. Because of the great importance of these data, molecular biologists have turned to computational scientists for help.

The history leading to this vast "accumulation of quantitative information" can only be sketched here. We could begin with the invention of the electron microscope in the 1940s. The electron microscope was the first instrument that could make observations within an order of magnitude or so of molecular resolution levels, thereby allowing researchers to see viruses for the first time. Viruses are generally too small to be observed with light microscopes.

Another invention that allowed researchers to begin to make observations at the atomic level is a technique called X-ray diffraction. X rays are electromagnetic radiation similar to visible light but with very short wavelengths, wavelengths that are roughly comparable to the size of an atom. Because of their short wavelengths, it might seem that X rays could be used to make microscopes that would be able to "see" atoms and molecules directly. But a batch of new technologies had to be developed to exploit this capability because conventional lenses and mirrors cannot be used to focus X rays as they do light rays. The main problem is that ordinary lenses and mirrors tend to absorb X rays rather than reflect or refract them. With the X-ray diffraction technique, X rays are scattered off a crystal made from the molecule of interest; the patterns made by the scattered X rays can be used to infer the locations of the atoms in the scattering crystal.

These technologies were critical in resolving a string of questions that go back to antiquity, as noted above, at least as far back as Aristotle's speculations about the process of heredity and the reproduction and growth of life-forms. But following Aristotle two millennia were to pass before Gregor Mendel worked out the basic rules of heredity in the 1860s. Mendel used careful quantitative experiments on the inherited characteristics of pea plants to discover that both parent plants possessed two copies of each of their genes, and that each parent contributed one of the two copies to each of their offspring. Each of the offspring plants therefore also had two copies of each gene, and the copies were selected randomly from the pair possessed by each parent. Mendel used the inherited colors in pea-plant flowers and other heritable characteristics to "see" this behavior of the inheritance of genes for the first time.

Mendel's groundbreaking work was ignored and forgotten in his own lifetime, but it was rediscovered around the turn of the twentieth century. This rediscovery of the laws of Mendelian genetics led naturally to a flurry of argument about the chemical nature of the Mendelian genes. The early consensus opinion was that the genetic information must be coded somehow into the protein molecules that are ubiquitous in every cell. But in 1943 Oswald Avery devised a set of classic experiments on inheritance in bacteria that resolved the question in a manner that surprised almost everyone. Avery established that it was not a protein but a hitherto obscure molecule called deoxyribonucleic acid (DNA) that carried the genetic information. Avery's discovery set off another flurry of activity to determine the molecular structure and behavior of DNA.

X-ray diffraction was the obvious tool to analyze the structure of DNA. But this application proved to be extraordinarily difficult because the DNA molecule was far larger than any molecule that had ever been imaged with X-ray techniques. It was not easy to make the necessary molecular crystals, but Rosalind Franklin worked through many of the difficulties and produced the first crude images of the structure of DNA in the early 1950s. Franklin's images were critical to the determination of the structure of DNA by James Watson and Francis Crick in 1953.

The importance of Watson and Crick's discovery can be explained briefly. The molecules that are involved in the vast majority of the chemical functions carried out by cells are entities called proteins, and proteins are simply linear chains of amino acids. There is an almost infinite number of possible amino acids that could be synthesized chemically, but only 20 of them are actually used to make protein molecules in all terrestrial life-forms. These 20 amino acids can be chained together in almost any order in strings, whose lengths range from a few dozen to a few thousand amino acids. A bewildering variety of protein molecules can therefore be constructed out of these twenty simple amino acid molecules. Hundreds of thousands of distinct protein molecules are found in the cells of the average human.

Watson and Crick's discovery showed how a cell "remembers" the correct sequence of amino acids in each of its protein chains, as well as how the cell manages to pass the information about the amino acid sequences to its "daughter" cells when it splits or divides in two. Their Nobel Prize–winning double-helix model for the structure of DNA looks something like a twisted stepladder. The vertical sides of the stepladder consist of chains of sugar molecules and phosphates, and the rungs of the ladder consist of pairs of molecules called "nucleotides." These nucleotides come in four different varieties that are abbreviated as A, C, G, T (technically adenine, cytosine, guanine, and thymine). The critical feature of the structure of DNA is that the pairs of nucleotides that constitute the "rungs" of the DNA stepladder are always constructed so that A is paired with T and C is paired with G. The pairs A-T and C-G are referred to as "complementary." Therefore, if the sequence of nucleotides along one side of the ladder is ACCATG, the complementary sequence along the other side must be TGGTAC. Watson and Crick noticed that this complementary feature would allow exact copies or replications of the DNA molecule to be constructed during the process of cell division by first splitting the molecule down the middle of each of the rungs of the DNA "stepladder" and then reassembling two new stepladders, one from each side of the original, matching the appropriate complementary nucleotides to each half of the original ladder.

George Gamow was the first to notice that in order to specify 20 different amino acids using only four nucleotides, they would have to be arranged in triplets. Clearly, you could not code 20 amino acids with only four nucleotides if they were used singly, and if you group the nucleotides by pairs, there are only 16 possible combinations. But there are 64 possible triplets, which is more than enough. Thus, the first question faced by researchers after Watson and Crick's determination of the overall structure of DNA was to crack this code and determine the correspondence between each triplet of nucleotides and its associated amino acid. (For technical reasons, the code is expressed as RNA triplets instead of DNA triplets. The information DNA is transcribed into RNA in the process of manufacturing proteins. The reason for specifying the RNA sequence is that RNA is single-stranded and does not have the ambiguity of complementary strands as in DNA. The only difference that we need to be concerned with here is that in RNA the nucleotide thymine is replaced by uracil, so T becomes U in RNA. The other three nucleotides are identical in DNA and RNA.)

The first and simplest of the code entries to be broken were triplets that used only a single nucleotide. It proved fairly easy to synthesize DNA molecules that consisted of only one nucleotide and its complement and then see which amino acid was found (exclusively) in the protein that was produced from this DNA string. Thus, AAA was found to code for an amino acid called lysine, and CCC was found to code for proline. Further decoding involved the careful construction of DNA chains that contained only one particular triplet of nucleotides, and by 1967, the DNA code was completely broken. Cracking the DNA code may have been one of the last major exploits in molecular biology that was done largely without computer power.

Because the matching of nucleotide triplets to amino acids is highly redundant (there are more triplets of nucleotides than there are amino acids), many different triplets were found to code for the same amino acid. And there is an interesting pattern to this redundancy. In many cases, the third nucleotide of the triplet makes no difference at all. Thus, the triplets GCU, GCC, GCA, and GCG all code for the amino acid called alanine. This has led some to suggest that the code was originally a doublet code and that only 16 amino acids or perhaps even fewer were originally involved, although no one has been able to devise a convincing mechanism for changing over from a doublet code to a triplet code. There are also three triplets (UAA, UAG, and UGA) that do not code for an amino acid at all but instead are sort of punctuation marks that indicate where a particular gene begins and ends in the DNA ladder.

After the brilliant work that was done to crack the DNA code, biologists were faced with the staggering task of trying to read the entire nucleotide sequence (or "genome") for various organisms and then interpret that sequence in terms of the functions and relationships of genes. The process of decoding the human and other genomes provides one of the best examples of the indispensable nature of computer technology in molecular biology. In this effort computer technology is paired with a string of other remarkable technologies including the polymerase chain reaction, gel electrophoresis, and restriction enzymes.

The first technology, the polymerase chain reaction, is needed simply to make multiple copies of a particular DNA molecule. This may not seem terribly important, but all the techniques that we presently have for reading a sequence of nucleotides in a DNA molecule involve destroying the molecule in a variety of clever ways. This is perhaps reminiscent of the techniques that physicists use to investigate the structure of subatomic particles by smashing them to bits, a technique that Feynman famously compared to investigating the structure and functioning of a pocket watch by hitting it with a sledgehammer. It is thus necessary to have a lot of identical copies of a DNA molecule in order to break it up in a variety of different ways to read its nucleotide sequence. The ability to replicate a particular DNA molecule is so important that it won separate Nobel prizes in chemistry for Arthur Kornberg and Kary Mullis. There is no fundamental reason that we could not develop techniques that would enable us to read a single intact DNA molecule the way a tape drive reads information from a magnetic tape; indeed, biological cells do this routinely. But at present we do not have any nondestructive techniques for reading the sequence of nucleotides in a DNA molecule.

The second technology, gel electrophoresis, is a fancy phrase for a technique that measures the length of a DNA molecule. Gel electrophoresis uses the force generated by an external electric field to drag an electrically charged DNA molecule through a gel. The speed of the molecule through the gel is a complicated but well-understood function of the electric charge and the friction or drag exerted by the gel, both of which depend on the length of the molecule. By measuring exactly how far the molecule is able to move through the gel during a standard time interval, researchers have a very sensitive technique for measuring the exact length of a DNA molecule. In other words, gel electrophoresis can give an exact count of the number of nucleotides in a short strand of DNA molecule. The ability to measure the length of a DNA strand is one of the critical technologies for the ability to decode the string. There are several techniques that exploit this capability; we will examine one of them to begin to understand the techniques involved.

This particular genome decoding technique is called the "chain terminator method" or sometimes the "dideoxy method" (see Voet and Voet, 1995, pp. 92–896). The chain terminator method begins by using restriction enzymes, which are proteins that have the ability to cut a double strand of DNA whenever they encounter a particular sequence of nucleotides. Each restriction enzyme has a particular sequence that it recognizes. This restriction enzyme is used to break a DNA chain into short segments. One of these segments is then replicated and split down the middle of the double chain DNA into its complementary single chains. Suppose that a single chain consists of the following very short sequence of nucleotides:

ATATGCTACAAC

We can use gel electrophoresis to determine that the chain contained 12 nucleotides. The copies of the single chain are then mixed in a solution that contains all four nucleotides together with an enzyme that can create a double strand DNA chain from a single chain by matching the nucleotides that are in the solution with their complementary nucleotides in the single chain. This enzyme can be compared to the talon of a zipper—it slides along the single chain and assembles the new double chain by attaching nucleotides as it moves along. By itself this is not very interesting. The operation of this enzyme would quickly reproduce lots of copies of the original double chain:

ATATGCTACAAC
TATACGATGTTG

But the chain terminator method has one more twist. Each nucleotide can be modified to create a version that has the useful property that when the modified nucleotide is included in the complementary chain then the enzyme that is creating the double chain terminates the process at that point. It therefore leaves only a truncated double DNA chain in the solution. (The modified nucleotides are called dideoxynucleotides because of the chemical nature of the modification, hence the alternate name for this method.) So, for example, a small amount of modified A, called A′, is included in the reaction solution that contains A, T, G, and C. But whenever an A′ nucleotide is randomly included in the complementary chain, the construction of the double chain would stop at that point. We would then find that after the reactions the following chains would be present in the reaction vessel:

AT
TA′

ATAT
TATA′

ATATGCT
TATACGA′

and

ATATGCTACAAC
TATACGATGTTG

We could then use gel electrophoresis to determine that we have chains of length 2, 4, 7, and 12, and from that we would know that A occurred at positions 2, 4, 7, and possibly 12 of the complementary chain. By repeating this process with the dideoxy-version of the other nucleotides, we can then tell the position of all four types of nucleotide in the chain. The nucleotide in the twelfth position on the chain could be determined from a knowledge of the behavior of the restriction enzyme that produced the fragment in the first place.

The chain terminator method is a marvelous technique for determining the sequence of short chains of DNA, but for technical reasons it is limited to chains no longer than about 800 nucleotides. And, of course, the human genome contains about three billion nucleotides. If the three billion nucleotide chain is broken completely into 800 nucleotide segments, then about four million segments will have to be sequenced. Fortunately the chain terminator method can be automated so that it can be run on computer-controlled machines. Therefore the process of sequencing four million segments need not exhaust the careers of hundreds of thousands of graduate students. This is one of the critical applications of computer technology to the process of sequencing a genome. But once the segments are sequenced individually they must then be reassembled to determine the complete original chain, and this reassembly process requires still more computer power.

In order to perform this reassembly and therefore sequence longer stretches of DNA, the DNA is broken randomly into subsegments and each of the subsegments is sequenced. Then the process is repeated with a different set of random breaks. In the terminology of the sequencing process, the genome is "oversampled," so that each part of the original genome is included in as many as 10 different subsegments that are sequenced. Then another critical application of computer technology comes into play. The decoded subsegments are matched up by computer programs that search for overlap sequences in the various subsegments, sequences that are identical in a particular pair of subsegments. The overlapping sequences then enable researchers to match up the subsegments and assemble then into longer strings.

In effect, the genome sequencers have created an enormous set of one-dimensional jigsaw puzzles that have to be assembled using computerized pattern-matching techniques. As we noted, the computer power required for this task is enormous, but it is well within current technological capabilities.

The complete process of decoding the genome is considerably more complicated than this abbreviated description; the February 15, 2001 issue of *Nature* and the February 16, 2001 issue of *Science* can be consulted for more details. One of the main complexities that has not yet been fully resolved relates to the fact that large segments of the genome do not seem to code for anything at all. They consist instead of millions of consecutive short repeat sequences such as AAGAAGAAGAAGAAGAAG . . . or CAGCAGCAGCAGCAGCAG. . . . These repeat sequences cause problems because, in the process of pattern-matching that is used to solve the great set of one-dimensional jigsaw puzzles of DNA chains, these repeat sequences can be matched almost anywhere that they exist. It is as though an ordinary jigsaw puzzle had been created by cutting a blank picture into perfect squares that can fit anywhere. These and a number of other problems have yet to be fully solved, and parts of the genome are not yet completely sequenced.

Letting computers handle this pattern-matching drudgery leaves far more time for researchers to ponder the really deep questions that are raised by the ability to see structures at the molecular level. These questions fall into a number of broad categories, and the solution of all of them will require massive amounts of computer power. As the International Human Genome Sequencing Consortium reported (2001, p. 860): "The human genome holds an extraordinary trove of information about human development, physiology, medicine and evolution. . . . Much work remains to be done to produce a complete finished sequence but the vast trove of information that has become available . . . allows a global perspective on the human genome."

This global perspective, the ability to "see" the genome, has already turned up a number of surprises that are likely to lead to major paradigm shifts in the future. Perhaps most surprising, the genome apparently contains only about 30,000 to 40,000 genes, only about one-third of early estimates. Indeed, this number is only about 50% larger than the genomes of worms and fruit flies. A related surprise is that only about 3% of the genome contains actual genes. The rest is referred to as "junk" DNA, such as the repetitive sequences mentioned above, with no known function. There are two broad schools of thought concerning this junk DNA—one proposes that it is really useless and is present in the genome because of evolutionary accidents. The other proposes that the junk DNA has some presently unknown function, that perhaps it is chemically necessary for organization of the very long molecule into tightly packed chromosomes inside the cell nucleus, for example. The argument that the junk DNA has some unknown function is undercut by the observed fact that the DNA of the pufferfish, which has also been sequenced, has almost no junk DNA, yet the pufferfish is a perfectly functional organism.

Another surprise found in the human genome is that mutation rates are about double in males as in females. This discovery has perhaps a natural evolutionary explanation: Males produce billions of gametes, while females produce only a few hundred in their lifespans. Therefore, harmful mutations in females would be much more likely to be passed along and result in harm to the offspring, whereas harmful mutations in males would merely kill off some fraction of the gametes or perhaps render them ineffective, leaving the remainder to proceed with the fertilization process.

Another surprise related to the "junk" DNA is the fact that in humans and most eukaryotes (organisms with nucleated cells) the sequences of nucleotides that code for a particular protein are not found in one continuous chain. Rather, they are in short sequences with batches of junk DNA interspersed. The interspersed pieces of junk DNA are called "introns." The DNA of bacteria (which do not have nucleated cells) is different in this regard. Bacteria generally do not contain junk DNA nor do they contain the same sort of introns as eukaryotes (although they do have some noncoding DNA of a different sort), and there is more nearly a one-to-one relationship between the bacterial DNA and the amino acids of the coded proteins. And some viruses such as bacteriophages ("bacteria-eaters") have an even stranger setup. Some of them are able to code for more proteins than they have DNA sequences. This particular puzzle was resolved when it was realized that the genes of the bacteriophage overlap each other in a peculiar fashion. The nucleotides are read in triplets, but if you move ahead in the string by a single nucleotide, you get the beginning of a totally different set of triplets. Thus, in our previous decoding example, the string ATATGCTACAAC could be read ATA, TGC, TAC, AAC, or, by skipping the first nucleotide: TAT, GCT, ACA, . . .

Computer analysis of genomes is expected to let us "see" aspects of the nucleotide strings that are of enormous importance to biology. For example, comparison of the genomes of different organisms will give new and important information about the phylogenetic relationships (the "family tree") of those organisms and their evolutionary history. The entire science of phylogeny, the study of the classification and evolutionary relationships of organisms, will be completely revolutionized. In the past, organisms had had their phylogenetic classification determined exclusively by their macroscopic physical characteristics: arthropods have external skeletons with jointed legs, annelids are wormlike with segmented bodies, chordates have a spinal cord and banded muscle structures, echinoderms have tube feet and radial symmetry, and so forth. But now we have a vast new suite of characteristics in gene structures that can be compared and contrasted using computerized comparison and analysis of their genomes. These new genomic characteristics may give clues to the deepest mystery in metazoan phylogeny, the relationships between groups at the phylum level (arthropods, chordates, and so forth). Naively, for example, one might expect chordates to be most closely related to annelids (segmented worms) because both have segmented bodies and bilateral symmetry. But biologists have long known from embryological studies that chordates are more closely related to echinoderms (starfish, urchins, and so forth) than to annelids. Genetic sequence data should soon help to clarify these and other similar relationships. Another case involves the microsporidian *Encephalitozoon cunculi,* an odd microscopic organism whose phylogenetic place was recently clarified by genetic sequencing techniques. *E. cunculi* is a eukaryotic (nucleated-cell) organism that lacks mitochondria, which are organelles that are present in nearly all eukaryotic organisms. When this was first discovered, it was thought that this lack of mitochondria implied that the microsporidians were very primitive eukaryotes, dating to a time when cell nuclei had developed, but mitochondria had not yet appeared. But a detailed analysis of the genome of *E. cunculi* indicates that they once had mitochondria and apparently lost them in the process of adapting to a parasitic lifestyle. They are apparently degenerate forms of fungus rather than primitive eukaryotes (Katinka et al., 2001). These and other evolutionary relationships are expected to be greatly clarified as genome sequence data from a much larger set of organisms becomes available.

One of the basic techniques for determining relationships among organisms and groups of organisms is a set of algorithms referred to as "cladistics." Cladistics was devised to determine relationships among populations of organisms according to the similarities and differences in their macroscopic characteristics. But cladistics can be equally well applied to their genomic characteristics, and there are expected to be vastly larger numbers of genomic characteristics that can be identified and used for this purpose. Cladistics was one of the most computer-intensive areas of biology, even when it was used with the smaller number of macroscopic characteristics; it will become even more computer intensive in the genome era. The task of sorting out the various combinations of genomic characteristics and understanding their phylogenetic relationships is one that would be completely impossible without the massive computation power of electronic computers.

Other major insights that have resulted from genome sequencing concern the very mechanisms that underlie the genetic variation that is essential to Darwinian evolution. One of the principal problems with genetic variation is that since genes specify the creation of proteins that presumably are essential to the functioning of an organism, then any change to a gene would potentially threaten the viability of that organism by eliminating an important protein. A possible solution to this problem was proposed by Susumu Ohno in 1970: If an organism were to duplicate its entire genome, then one copy would be available to maintain the synthesis of essential proteins, but the other copy would be available for random experimentation. Most of those random experiments would not produce useful proteins but still would not harm the organism because it had another back-up copy of the gene that was still functioning. At the time that Ohno made this suggestion there was no way to check it directly, but with modern gene sequencing techniques the doubled genome would be fairly obvious. And recent gene sequencing results appear to confirm the idea (Pennisi, 2001).

Many biologists feel that the natural next step beyond sequencing the genome will be to develop an understanding of the structure and function of the proteins whose amino acid sequences are coded in the genome. Indeed, a new word, the "proteome," has been coined to describe the entire set of proteins whose code is contained in a particular genome. And studying the proteome promises to involve perhaps many hundreds of problems, each of which may require as much computer power as the entire genome sequencing project.

At first glance, the problem of studying the proteome appears very similar to the genome. DNA is a linear string of nucleotides that have been sequenced, and proteins are linear strings of amino acids whose sequence is known if the DNA nucleotide sequence of their respective genes is known. But the major difference between them is that DNA is functional in its linear form while proteins are not. The linear protein molecules immediately fold themselves into a bewildering variety of physical shapes, and those physical shapes are of critical importance to their biological functioning. Biologists have known for some years that proteins can be unfolded to their linear (and nonfunctional or denatured) shape by simply heating them. When they are cooled they can refold into their often bizarre-looking but functionally correct form. The problem of calculating how a protein should fold up may appear to be a simple problem—indeed, the molecule accomplishes it effortlessly, and, one might say, brainlessly. But the number of possible folding configurations is so large that no one has yet figured out an effective method for calculating these folding properties, that is, a computational method that is within the capabilities of present computers. The calculation may, in fact, require more computer power than exists today. This, of course, only means that we have to wait a few years for the necessary computational capability that doubles every year or two according to Moore's famous law.

Today proteomics is at a primitive state compared to genomics. Service (2001, p. 2074) wrote:

> In terms of complexity, proteomics makes genomics look like child's play. Instead of an estimated 30,000 to 40,000 genes, protein experts think that humans have somewhere between 200,000 and 2 million proteins. What's more, whereas genes remain essentially unchanged through life, proteins are constantly changing, depending on the tissues they're in, a person's age, and even what someone ate for breakfast.

And Sydney Brenner, one of the pioneers of the Human Genome Project, put it this way: "You can take DNA from anything—yourself, bananas, barnacles, and put it through a machine. That's because it's all the same stuff. [Conversely] there are no good techniques to try to handle proteins" (quoted in J. Cohen, 2001, p. 57). Cohen goes on to explain some of the complexities of protein functions:

> Proteins provide the structure of all cells and allow them to move around. They make up the cacophony of messengers that constantly traffic between immune-system cells. . . . They control the firing of neurotransmitters that allows us to think, the contraction of muscles that allows us to move, and the very on-off switches in our genes that allow us to make even more proteins. Proteins blow genes out of the water in sheer numbers, too.

Yet computer technology is beginning to make inroads into the vast difficulties of determining protein structures and functions. Service (2001, p. 2074) describes a modern computerized laboratory that is designed to identify proteins:

> In a labyrinth of rooms . . . four different kinds of bench-top robots—24 machines in all—steadily work together in silence. The robots, some weighing as much as 150 kilograms with arms that whir in all directions, carefully isolate a mix of proteins from a tissue sample, separate them into clumps of identical proteins, chop members of each clump into fragments, and place them into an array of wells on tiny metal plates. Technicians feed these plates into a series of 51 mass spectrometers worth over $150,000 each; every second, each of these refrigerator-sized machines spits out a fingerprint of a protein fragment based on its mass. A supercomputer then compares each fingerprint to a database to identify the amino acids it contains. Then, within minutes, it reassembles the jumble of fragments to identify the proteins from which they came. The result: a list of thousands of proteins present in the starting sample. A few years ago, identifying just one of these proteins often took years. Today it takes hours [to identify thousands of them].

As Brenner noted, there is no single approach that can identify all possible proteins. Isotope-tagging and electrophoresis are useful tools, and some researchers are attempting to make "protein chips," specially designed arrays of molecules that capture specific molecules at specific sites. This approach was effective in genome studies because the DNA molecule has two complementary chains. Therefore it is straightforward to design a molecule that will capture any given chain—you merely synthesize its complement. Proteins, in contrast, tend to be highly specialized in the types of molecules that they will bind to, and designing the molecules that can be used on protein chips is a much more difficult problem than the comparable problem in genomics.

The human genome data has made protein identification much more simple. Many proteins can be identified by sequencing a small segment of the protein, then searching the genome data for a nucleotide sequence that would produce that fragment of protein. If a match is found, the genome sequence can immediately give you the rest of the protein sequence. Of course, there is a chance that two proteins have identical subsequences, but this problem can be handled by making the subsequence long enough and by internal consistency checks—for example, the gene has to code for a protein of the correct weight.

Clearly, understanding the structure and functions of proteins at the atomic level is of central importance to molecular biology. And computers provide the only hope for dealing with the immense quantities of information needed for further work in proteomics. There are vast sets of new problems that need to be solved, and work on them is only beginning. But the work gives some hope of providing for the first time a detailed understanding of the functions of cells at the molecular level. And naturally this detailed understanding can be expected to set off new paradigm shifts.

Although the problem of sequencing genomes is essentially solved, there remains an enormous amount of work to be done, millions of different species whose genomes can be sequenced. And determining the proteome is a much more difficult problem, one that no one would claim is solved. But there is yet another level of problems that are essential to biology. Once the genome is completely known, and the proteome is known, there remains the incredibly complex problem of understanding how all of the molecular components interact in biological systems. For proteins alone, if there are 500,000 proteins, then there are potentially 500,000 squared ways (about 250 billion ways) that proteins could interact pairwise. And proteins tend to interact in more complex ways than just pairwise. Critical chemical reactions in organisms tend to involve chains of protein reactions. So there is a nearly unlimited set of problems to be worked out after the genome and the proteome are understood, and that set of problems involves just a single organism. Therefore, although computers have provided unprecedented power that is producing the largest phase change in the history of biology, there is no chance of exhausting the problems of biology in the foreseeable future.

The phase change associated with the new ability to see biological structures at the atomic and molecular levels is already producing some of the major paradigm shifts of modern biology, and we can confidently say that it will produce more. It is also changing the basic research methods of biologists in ways that will require vast amounts of computer power. As Butler (2001, p. 768) put it:

> Many research leaders predict that the potential to integrate different levels of genomic data . . . will radically change biological research. . . . Some foresee an era of "systems biology," in which the ability to create mathematical models describing the function of networks of genes and proteins is just as important as traditional lab skills. . . . "Every institution that expects to be competitive in this new era will need to have strengths in high-throughput computational approaches to biology" says Francis Collins, director of the National Human Genome Research Institute.

Biology is therefore very similar to the other sciences that we will examine in later chapters in this book: It is characterized by phase changes in the past that resulted from newfound abilities to see things, such as the phase change that followed the invention of the microscope, and this phase change touched off a string of paradigm shifts. And the computer is not only analogous to the invention of the microscope in enabling us to see things at the molecular level that could not be seen without the computer, but it is also essential to handling the flood of data that has resulted from our newfound ability to see things at the molecular level. This combination of capabilities, both to see things and to analyze the flood of data that results from seeing new things, may be unique in the history of biology, but it is characteristic of the impact of computers in many other fields of research.

4

PHASE CHANGES IN PHYSICS

Astronomy and biology were easier areas in which to explore phase changes because there is a broad general familiarity with telescopes and microscopes and with the capabilities of these instruments. Physics and mathematics, by contrast, are fields that many find far less familiar, fields that may seem much more esoteric and complicated. For example, many people who have a basic understanding of the working of a telescope or a microscope might have some difficulty describing the operation of the Tevatron accelerator at the Fermilab in Illinois. Yet we will find that the familiar patterns that were seen in astronomy and biology occur in physics as well. Historical phase changes in physics are marked by the invention of novel ways to see and detect things that could not be seen prior to the phase change. And, just as in other areas, these phase changes touched off paradigm shifts when the things that were newly seen did not fit well with earlier paradigms.

Physicists' abilities to see and detect objects have expanded immensely over the last century, especially objects at atomic and subatomic levels. It is difficult to remember how very limited those abilities were back at the turn of the twentieth century. Today no one would seriously question our knowledge of atoms and their general properties and structures, but a hundred years ago even the very existence of atoms was still a matter of some controversy. Physicists of the caliber of Ernst Mach had argued against the existence of atoms largely because they could not be seen or detected with any known technology. The idea that atoms not only exist but also that they have internal structures that can be changed and transmuted was too outrageous to even be considered. No less a theorist than James Clerk Maxwell wrote in 1875: "The formation of the atom is therefore an event not belonging to that order of nature under which we live. It is an operation of a kind which is not, as far as we are aware, going on on Earth, or in the sun or the stars, either now or since those bodies began to be formed. . . . [Until] the very order of nature itself is dissolved, we have no reason to expect the occurrence of any operation of a similar kind" (quoted in Andrade, 1964, p. 73). One of the important phase changes that occurred in physics in the early twentieth century centered on the development of newfound abilities to see and manipulate things at the atomic and subatomic level. This phase change caused paradigm shifts in our ideas about atoms, not to mention their internal structures and the forces that maintain them.

The invention of photography was perhaps even more important for this phase change in physics than it was for astronomy. The ability to see things at subatomic levels began with Henri Becquerel's celebrated use of photographic plates to "see" the effects of the natural radiation generated by decaying uranium atoms. In 1896 Becquerel discovered that uranium compounds could fog nearby photographic plates even when those plates were sealed into light-tight containers. Ernest Rutherford went on to investigate the nature of the phenomenon that Becquerel had discovered. He came to the stunning conclusion that some atoms were unstable, that they were being transmuted by natural decay processes. These atomic decays gave off energetic particles that were able to fog photographic plates.

In 1897, J. J. Thomson began a series of experiments to measure the mass of the electron; he found it to be about three orders of magnitude smaller than the mass of a hydrogen atom, the lightest atom known. Thomson's measurement provided the first serious evidence of the existence of things that are much smaller than atoms. Thomson made the further suggestion that these electrons might be components of a structure that exists inside atoms. He proposed the "plum pudding" model for the structure of atoms, that they consist of electrically neutral combinations of positively and negatively charged particles that were held together by their mutual electrostatic attraction in a sort of subatomic pudding.

The phase change that is marked by physicists' ability to see things at subatomic levels continued with Rutherford's use of alpha particle radiation to investigate Thomson's plum pudding model. Alpha particles are one of the components of the radiation emitted from decaying uranium atoms; today they are recognized as highly energetic nuclei of helium atoms. Rutherford adopted a technique that had been invented by Crookes that used phosphorescent screens of zinc sulphide to detect or see these alpha particles. When an alpha particle strikes the zinc sulphide, it gives off a faint flash of light that can be observed with a low-power microscope. Rutherford first discovered that when you place a thin sheet of a material such as mica into a beam of alpha particles, the beam would penetrate the sheet but it would become blurred or defocused. Apparently the atoms in the mica deflected the alpha particles slightly but did not stop them.

Rutherford and his colleagues Hans Geiger (who was later famous for inventing the Geiger counter) and Ernest Marsden then investigated this defocusing of the alpha particle beam by placing a thin sheet of gold foil in the beam. The thickness of the foil could be carefully controlled, and it contained only a single type of atom. In 1909, Marsden observed that although most of the alpha particles passed straight through the gold foil with only slight deflections, some of them were reflected almost straight backward. This behavior baffled Rutherford for a time, but he eventually came to realize that he could explain the observation only if the plum pudding model of the atom were wrong, and that instead most of the volume of the atom consists of empty space. The data made sense if nearly all the mass of the atom and all its positive electric charge were both concentrated in a tiny body or nucleus at the center of the atom. The alpha particles would bounce backward when they fortuitously made a direct hit on this tiny nucleus; otherwise, they would be only slightly deflected. This discovery of the atomic nucleus led to what was called the "solar system" model for the structure of the atom. The nucleus is analogous to the sun at the center and Thomson's electrons orbit about the nucleus in a manner analogous to the planets.

Other physicists soon followed up on Rutherford's brilliant insight, that beams of energetic subatomic particles could be used to "see" structures inside atoms and eventually to see structures inside things that were even smaller than atoms. Rutherford had been able to use the natural decay of unstable atoms to create his probing beams of particles, but it was soon realized that this natural radioactivity had serious limitations—limits on both the types of particles available, and, more seriously, limits on the amount of energy in the beam of particles. For technical reasons in quantum mechanics, observing smaller and smaller particles required larger and larger beam energies.

A concerted effort was made to develop methods to generate and control beams of very energetic subatomic particles. Van de Graaff and Cockroft and Walton devised ways to accelerate such particle beams using enormous electric fields (high voltages) that would push particles to very high energies. In effect, these researchers generated enormous "ski runs," using electric fields to accelerate charged particles just as a gravitational field accelerates a skier as he goes downhill. The charged particles would run "downhill" through a strong electric field. These techniques soon ran into limits on the sizes of the machines and the magnitude of the voltages that could be sustained without accidentally generating artificial lightning bolts. Indeed, Van de Graaff's generator is still used to generate small lightning bolts in college laboratories.

A major advance was made by E. O. Lawrence with the invention of the cyclotron. Continuing the ski hill analogy, Lawrence's invention operated not by making the hill higher and higher (i.e., larger and larger voltages) but by using a small hill, a small voltage, repeatedly. In Lawrence's apparatus, as soon as the particle gained energy by running down a small hill, the voltage was changed so that the particle found itself suddenly back up at the top of another small hill and ready to run down again. Lawrence's cyclotron was able to repeat this process over and over to generate extremely energetic subatomic particles without needing extreme voltages.

These particle accelerators, sometimes popularly called "atom smashers," were critical to a sequence of important paradigm shifts in physics. The improved ability to "see" structures at the subatomic scale led to the discovery of many things that did not fit earlier paradigms, including such things as protons, neutrons, mesons, and antimatter (some of which were actually discovered first in natural cosmic radiation, but whose properties were explored using particle accelerator technology). The data taken with successively more powerful accelerators provided a wealth of information about the forces and particles found inside atoms and their nuclei. In particular, two new forces, called strong and weak, were found to operate inside the atomic nucleus. These discoveries were exploited by generations of theorists, including Bohr, de Broglie, Schrodinger, Heisenberg, Dirac, Feynman, Gell-Mann, Weinberg, and others too numerous to list. Their work culminated in the development of what is now called the "standard model" of particle physics that provides the theoretical basis of our understanding of the behavior of all the known particles and forces in physics, with the notable exception of the force of gravity. But that is getting a little ahead of our story.

Just as with the development of large telescopes, the construction of larger and larger particle accelerators ran into serious difficulties in the precomputer era. One of the critical problems was that the beams of subatomic particles had always been aimed at fixed, unmoving targets. In Rutherford's case the fixed targets were the gold atoms in his sheets of foil. The use of fixed targets did not cause serious problems for relatively low-energy experiments, but as the energy of the beam grew larger, a greater fraction of the energy of the beam was dissipated in simply accelerating the target atoms, merely pushing them to high speeds rather than producing interesting physical reactions.

The solution to this problem lay in the use of pairs of particle beams that would collide head-on. When a particle in one of the beams strikes one from another beam that has the same energy but is moving in the opposite direction, not only is the interaction energy doubled over the energy of a single beam, but also, much more important, no energy at all is dissipated in simply accelerating the atoms inside a target mass. And the operation of these colliding beam accelerators requires computer technology. In fact, when the idea of colliding beams was first proposed, many skeptics claimed that although the idea looked very good in principle it could never be practical because the probability of getting a head-on collision was too small to be of any use. Glancing collisions would be nearly useless—for practical work you would need large numbers of head-on collisions.

To get large numbers of head-on collisions you need to improve what is called the "luminosity" of the beam, which is a fancy way of saying that you need more beam particles concentrated into a smaller space in order to increase the probability of a collision when the particle beams intersect. But there are a number of problems associated with improving the luminosity of a particle beam. There are a lot of different effects that tend to defocus or spread the beam out, decreasing its luminosity. These effects include collisions with residual atoms of gas in the vacuum beam chamber, irregularities in the magnetic fields that focus the beams, even the mutual repulsion of the electric charges of the beam particles themselves.

The luminosity of the beam can be improved by adjusting some controlling magnets whose special magnetic fields are able to focus the beam into a smaller volume. But in order to know exactly how the magnets should be adjusted, you first have to measure the deviation of the beam away from its optimum focused condition. Then you have to get that deviation information to the controlling magnets farther down the accelerator, ahead of the beam. Since the particle beam travels at nearly the speed of light, and information cannot be transmitted faster than lightspeed, it would seem that the critical information can never catch up with the beam in order to focus it. But in most modern accelerators, at least since Lawrence's original invention of the cyclotron, the accelerator beams are bent into a circular shape so that the same focusing magnets and accelerating fields can be used over and over again on the beam as it traverses a circle at nearly the speed of light. Simon van der Meer won a Nobel Prize for developing a practical scheme that transmits the focusing information across the diameter of the circular accelerator faster than the beam is able to travel around its circumference.

Naturally, getting the critical information across the accelerator and adjusting the focusing magnets before the beam travels around half the circumference requires the speed and information-processing capabilities of computer technology. And this use of computer technology to make colliding beam accelerators practical gives us a simple, somewhat unsophisticated way to measure the magnitude of the phase change associated with computer technology and compare it to the magnitude of the earlier phase change associated with particle accelerators in the precomputer era. As we noted above, the problem with using a powerful beam on a fixed target is that most of the beam energy simply causes the target atoms to move, and that is not a terribly interesting thing to do. The energy that is left for interesting reactions is simply the square root of the beam energy. But in a collider the available energy is twice the beam energy, because the full energy of two colliding beams is available. The most powerful accelerator under construction today is CERN's (European Organization for Nuclear Research) Large Hadron Collider (LHC), which is designed to produce beams with an energy of 7 trillion electron volts. If the beams were used against fixed targets, only about 2.6 million electron volts would be available (the square root of 7 trillion). But in the colliding beam mode, 14 trillion electron volts are available, an improvement of more than six orders of magnitude over what is practical without computer technology (and that is assuming that the beams could be operated in fixed-target mode without computers).

The use of computers to make colliding beams practical is only one component of the use of computers in the LHC. The operation of modern colliding accelerators creates problems for physicists very similar to the problems created for biologists by the decoding of the genome. In both cases the amount of information generated is enormous, far larger than can be handled by any reasonable amount of human effort in the absence of computers. The computer-enhanced luminosity of the LHC, for example, is expected to produce nearly one billion particle collisions per second.

Not all of the billion collisions produced each second in the LHC will be intrinsically interesting; many of them will only generate subatomic particles whose properties are already well known and studied. Only about 10 out of each billion collisions are expected to produce new and interesting physics, and only these 10 will be recorded and kept for further analysis. Of course, computers are needed to sort through the billion collisions and decide which 10 or so should be saved and then repeat this process in the following second. Clearly, this task is completely impossible in the absence of computer technology. And even though the computers will save the information from only about 10 collisions in each second, this is still a staggering amount of information, about a million collisions per day. No unaided human analyst could deal with collisions at this rate.

Contrast the event rate of a billion collisions per second in the LHC to the rates that Rutherford was able to attain in the precomputer era using the naked eye to detect and count the phosphorescent flashes produced by alpha particle beams. Rutherford's techniques ran into serious difficulty whenever the count rate became much higher than one particle per second, as fast as the human observer could count them and record the count by hand. And the events that Rutherford observed were far less complicated than the reactions that the collisions in the LHC will produce. So again, as with the astronomer's need to find rare events such as the occultation of a distant star by a distant asteroid in chapter 2, the ability to analyze the data from vast quantities of collision events gives researchers the ability to find rare occurrences, unusual subatomic reactions and reaction products (new subatomic particles) that would not otherwise be seen. Therefore, another way to compare the magnitude of the phase change produced by present-day computer technology to the phase change that was caused by Rutherford's initial use of particle beams to probe subatomic structures is simply to look at the ratio of the count rates that are attainable with and without computer technology, about nine orders of magnitude.

There is another paradigm shift ongoing in physics today, potentially one of the greatest conceptual revolutions in the entire history of our understanding of the universe. Physicists today are seriously exploring the possibility that the universe may have more dimensions than were previously suspected. Instead of the usual three dimensions of space, or the four dimensions of space-time, it now appears that the universe may have 11 dimensions, 10 of space and one of time. To understand this new paradigm shift we have to examine the phase change that preceded it, a phase change that allowed us to see a variety of new phenomena that we now think could demand the existence of 11 dimensions. This phase change can be traced back more than two centuries.

The story begins in 1791, with a simple observation by Luigi Galvani in Italy. Galvani noticed that if he connected two different metals—such as copper and iron—together and then touched them to a dead frog's leg, the leg would twitch. This was an observation that anyone could have made as far back as perhaps the early Iron Age, but as far as we know no one else ever did. Galvani had produced an electric current in the laboratory using what we now call the junction potential between two metals. This junction potential is the basis of battery technology to this day.

Alessandro Volta built on Galvani's observation and created a stack of metal plates or an "electric pile" that was the first practical electric battery. Volta's battery vastly expanded the ability of researchers to see, observe, and control the effects of electric currents and voltages in the laboratory. No longer were they limited to age-old techniques such as generating static electric charges by rubbing amber with cat fur.

Electricity and magnetism had both been known from antiquity, and it had always seemed that the two phenomena should be related in some way—static electricity and lodestones were both known to exert a mysterious force on other objects. But the relation between electricity and magnetism remained obscure until the invention of the battery allowed researchers to experiment with electric currents. Volta's battery created a phase change in physics by allowing researchers to see and manipulate the effects of electric currents. And in 1820 the Danish physicist, Hans Oersted, made a critical breakthrough when he saw that although static electricity had no obvious relation to magnetic effects, an electric current consisting of electric charges in motion would produce a magnetic field perpendicular to the current. Oersted's simple observation was followed up by a number of prominent scientists including Ampere, Faraday, and Henry, who developed the vital insight that a changing magnetic field induces (generates) an electric field, and a changing electric field induces a magnetic field.

This seemingly innocuous fact would lead eventually to the first major shift in our understanding of the dimensionality of space-time. The next important step in the story belongs to James Clerk Maxwell, who developed a theoretical model that expressed everything that was known about electric and magnetic phenomena as a single set of differential equations. Maxwell's justly celebrated set of equations produced one of the most important paradigm shifts in the entire history of physics, a theoretical breakthrough on a par with Isaac Newton's discovery of the inverse-square law of gravity. Maxwell was able to show that his equations had a solution in the form of a wave of electromagnetic fields. In Maxwell's electromagnetic waves, a changing (oscillating) electric field induces a changing magnetic field, which in turn induces a changing electric field and so forth as the wave propagates through space. Maxwell calculated that the propagation speed of this wave would be the same as the speed of light to the accuracy that it was then known, and he correctly proposed that light itself was a particular frequency band of these electromagnetic waves. He further predicted that electromagnetic waves should exist with other frequencies. In other words, there were previously unknown forms of electromagnetic radiation, some of them with lower frequencies than visible light (infra-red light and radio waves) and some with higher frequencies (ultra-violet light, X rays, and gamma rays). These new frequencies of "light" were invisible to the naked eye because the eye is sensitive only to a very narrow frequency band.

Invisible light was a genuinely revolutionary idea, a paradigm shift that underlies much of modern communication technology. Maxwell's prediction was brilliantly confirmed in 1889, when Heinrich Hertz produced and detected the first artificially generated radio waves (lower frequency electromagnetic radiation). The realization that light is an electromagnetic phenomenon, and that other previously unsuspected forms of electromagnetic radiation exist with frequencies that are invisible to the human eye, was perhaps the first major paradigm shift that resulted from the phase change that followed Volta's invention of the electric battery.

Maxwell's electromagnetic theory is sufficiently accurate that it remains the basis of much of the electronics industry today, but it had two rather subtle problems that were quickly recognized. One of them was discovered when Lord Rayleigh and Sir James Jeans tried to use Maxwell's equations to calculate the spectrum of electromagnetic radiation that would be generated by a warm body. Rayleigh and Jeans discovered that according to Maxwellian theory any warm body should radiate enormous, essentially infinite quantities of energy at very short wavelengths, the so-called ultraviolet catastrophe. But no such short wavelength radiation was actually observed. This observation led Max Planck to develop the earliest version of quantum theory.

Planck attempted to fix the error in Rayleigh and Jeans's calculations by using the newly invented concepts of statistical mechanics that had been developed partly by Maxwell himself. Statistical mechanics is a technique that uses the methods of mathematical statistics to calculate the macroscopic (large- scale) properties of a large ensemble of small interacting bodies or particles such as the atoms in a sample of gas. Maxwell used statistical mechanics to correctly derive the velocity spectrum or distribution of velocities of molecules in a warm gas. This Maxwellian distribution of gas molecule velocities was similar in shape to the observed spectrum of the radiation of a warm body, and this similarity led Planck to attempt to use the same mathematical techniques to try to resolve the problem of Rayleigh and Jeans's "ultraviolet catastrophe."

To use the ideas of statistical mechanics Planck had to assume that electromagnetic radiation occurred in "packets" analogous to the atoms of gases in Maxwellian statistical mechanics. He originally did not intend these packets to be anything more than a calculational convenience that would be eliminated by evaluating the theory at the mathematical limit where the size of the radiation packets approached zero and the packets disappeared. However, when Planck did that calculation, he obtained results identical to those of Rayleigh and Jeans, another ultraviolet catastrophe. He then won a Nobel Prize when he discovered that his theory provided good agreement to the observed electromagnetic spectrum of warm bodies if the size of the packets of radiation did not approach zero but instead remained small and nonzero. A few years later Einstein also won a Nobel Prize by showing that Planck's assumption that electromagnetic radiation occurred in packets or quanta also provided an explanation for the way in which electrons were observed to be kicked out of metals by electromagnetic radiation, the so-called photoelectric effect.

This discovery that electromagnetic radiation was quantized in packets (packets that were later called photons) was the second major paradigm shift to follow from the phase change produced by Volta's electric battery. And quantum theory, when combined with data from particle accelerator experiments that were noted above, culminated in the development of the standard model that unified our theoretical understanding of all the fundamental forces of physics, except gravity. But until very recently, every attempt to reconcile gravity (Einstein's general relativity) with the standard model met with conspicuous failure.

This difficulty in reconciling quantum mechanics and relativity theory leads us to examine the second subtle problem that had been found in Maxwell's electromagnetic theory. The problem is related to an essential feature of Newtonian mechanics. In Newtonian theory, all inertial (nonaccelerated) frames of reference are equivalent. Thus, if you set up a laboratory inside a closed railroad boxcar, there is no way in Newtonian theory to determine whether the boxcar is sitting idle on a side track or traveling down straight tracks with a uniform velocity. But Maxwell's theory behaved differently inside the boxcar laboratory. As Greene (1998, p. 5) put it: "According to Isaac Newton's laws of motion, if you run fast enough you can catch up with a departing beam of light, whereas according to James Clerk Maxwell's laws of electromagnetism, you can't."

The conceptual problem, as Einstein expressed it, is perhaps easiest to understand when the laboratory inside a closed boxcar moves at the speed of light. If you observe one of Maxwell's electromagnetic waves moving in the same direction as the boxcar (and at the same speed, of course), then using Newtonian concepts you should expect to see a wave that is stationary (standing still) relative to the laboratory. But a stationary electromagnetic wave is not a solution to Maxwell's equations. The observed wave would consist of a set of stationary electric and magnetic fields without any electric (or magnetic) charges to maintain those static fields, and there is no such solution to Maxwell's equations. In Maxwell's theory, the electric and magnetic fields both have to change with time in order to induce the other field, and the necessary time changes to the fields would not be observed in the stationary wave that you might expect to see inside the moving boxcar.

There were a number of possible ways to try to resolve this difficulty. Albert Michelson devised an extraordinarily sensitive experiment to measure the effect of a uniform laboratory velocity on electromagnetic radiation. The Michelson-Morley experiment was famously unsuccessful—it found no measurable effect of the motion of the laboratory. A number of physicists took a different tack and tried to modify Maxwell's equations to make them consistent with the behavior of Newtonian physics. They also failed. It was Einstein who made the critical breakthrough. He realized that if Maxwell's equations were to be preserved and still be compatible with Newtonian mechanics, then our concepts of space and time would have to change. In particular, we could no longer regard the universe as consisting of three dimensions of absolute space and one dimension of absolute time. Following up on earlier ideas by Fitzgerald, Lorenz, and others, Einstein showed that to preserve both Maxwell's equations and Newtonian dynamics in arbitrary inertial reference frames, the length of a body in the direction of the laboratory's motion would have to decrease (the "Fitzgerald contraction"), and the length in the time direction would have to increase.

These contractions and time dilations constitute one of the most famous paradigm shifts in modern physics and are the basis of the first change in our concepts of the dimensionality of the universe since Euclid's time. In Einstein's theory, space and time must be treated as a four-dimensional continuum. We can try to develop a conceptual understanding of why this is true by making an analogy between Einstein's relativistic transformations of space and time coordinates and a transformation that is more familiar, a simple rotation. The relativistic transformations are, in fact, slightly more complicated than simple rotations, but the idea of a rotation conveys many of the critical ideas.

To understand the analogy, suppose you were to construct a model city with straight streets that run perpendicular to each other and you place the model on a turntable so that it can be rotated. Initially, you place the model so that one set of streets runs directly in the north-south direction and the cross-streets run in the east-west direction. You can then rotate the turntable by a small angle. The streets that originally ran north-south will now run diagonally, partly north-south and partly east-west. You can even turn the table by 90 degrees so that the streets that originally ran north-south now run east-west.

These rotations of the model city are similar to what happens in Einstein's theory, but with the critical difference that the rotations can occur through time as well as through space. Thus, a spatial direction appears to be squeezed and a time dimension appears to be stretched because a spatial coordinate has been rotated into a time coordinate.

These rotations through the direction of time constitute the main reason that time must be considered as a fourth dimension in relativity theory on a par with the other three dimensions. It is more than merely a matter of accounting convenience to list the time coordinate along with spatial coordinates. Relativity cannot work if time is considered separate from and independent of spatial dimensions as it is in classical Newtonian theory. The necessary transformations of relativity theory require that time be convertible into space and vice versa. As we will touch on later, there are ways to use gravity to transform a spacelike coordinate entirely into a timelike coordinate, analogous perhaps to rotating the model city by 90 degrees across time as well as space. The requirement that transformations be made across both spatial and temporal directions was the third major paradigm shift that was a direct and forced result of the phase change that resulted from the ability to see and manipulate electric currents in the laboratory. In other words, without Volta's invention of the electric battery or something comparable to it, there would have been no Einsteinian paradigm shift.

Einstein went on to develop what is called "general theory of relativity" (a theory of gravity). This theory showed that gravitational forces could be understood as a manifestation of the bending or curvature of four-dimensional space-time. The idea that space and time could be curved and must be described in terms of non-Euclidean (curved-space) geometry was yet another major paradigm shift that resulted from the need to reconcile Maxwell's equations with Newtonian mechanics. And the idea that space-time can be curved or "bent" strongly suggests the possibility of still more dimensions of space-time. Additional dimensions would provide enough room for a four-dimensional space-time to bend or curve. There is, in fact, a theorem in differential geometry called the "embedding theorem," which states that a curved N-dimensional space can always be embedded in an uncurved or "Euclidean" space whose number of dimensions is larger than N. This sounds esoteric, but it can be illustrated with a familiar example. The curved two-dimensional surface of a sphere cannot be embedded in a two-dimensional Euclidean (uncurved or flat) plane, but it can be embedded in a three-dimensional Euclidean space, as it is in our ordinary conception of the surface of a sphere in three-dimensional space. Thus, the curvature of an N-dimensional space always suggests the existence of a higher-dimensional space that might be uncurved. (The higher dimensional space could itself be curved, implying yet more dimensions.) Because Einstein's theory of gravity (general relativity) required the curvature of four-dimensional space-time, it therefore contained the first serious suggestion that our universe may contain more than four dimensions.

The next clue that four dimensions were not enough for space-time, and that still more dimensions would be needed to make sense out of the forces that physicists had discovered and were trying to understand, came almost immediately in the long-unappreciated work of Theodor Kaluza. In 1919 Kaluza showed that electromagnetic forces as well as gravitational forces could be explained if you assume that the universe consists of a curved five-dimensional space-time continuum. Oscar Klein expanded on Kaluza's brilliant insight, but the work was not pursued further at that time, in part because it was nearly a century ahead of its time. There were other forces that had to be understood and considerably more mathematics that had to be developed before significant progress could be made in understanding the dimensionality of space-time.

In the early part of the twentieth century, physicists knew of the existence of only two forces—Newton's gravity and Maxwell's electromagnetism. But over the next few decades a series of experiments with both natural and artificially accelerated particle beams allowed researchers to see the effects of two more forces. These two forces were called strong and weak and have effects that are largely confined inside the atomic nucleus. In 1979, Sheldon Glashow, Abdus Salam, and Steven Weinberg won a Nobel Prize for developing a model that unified the weak force with electromagnetism, and shortly thereafter physicists developed a theoretical formalism, the so-called standard model mentioned above, which unified the strong, weak, and electromagnetic forces in much the same way that Maxwell's theory had unified electric and magnetic forces. Indeed, the standard model and Maxwell's theory have a great deal in common including a vital but mathematically somewhat esoteric property known as "gauge-invariance." But there was a serious problem with the standard model: it did not include gravity. Worse, every attempt to include Einstein's gravity theory (general relativity) had failed spectacularly. As Greene put it, "As they are currently formulated, general relativity and quantum mechanics [as exemplified in the standard model] *cannot both be right*" (Greene, 1998, p. 3).

Before proceeding with recent efforts to unify the standard model with general relativity, it might be useful to summarize the somewhat complex developments that have brought us to this point. The work of Becquerel, Rutherford, and others initiated a phase change by developing techniques for observing structures at subatomic levels using beams of energetic subatomic particles. Two previously unknown forces, the strong and weak nuclear forces, were discovered when their effects were seen for the first time. Additional phase changes accompanied the use of computer technology to make colliding beam accelerators practical, and the use of computer technology to record and analyze the data from these colliding beams. The measurements made possible by these phase changes revealed the existence of unsuspected levels of structures inside atoms and subatomic particles. A sequence of paradigm shifts resulted in the formulation of the standard model that provides a unified theory of electromagnetism and the strong and weak forces. Meanwhile, another phase change was generated by the ability to generate and observe the effects of electric currents in the laboratory. This phase change led to a sequence of paradigm shifts, starting with the development of Maxwell's equations for electrodynamics. Then the attempt to reconcile Maxwell's equations with statistical mechanics and the spectrum of radiation of warm bodies led to Planck and Einstein's discovery of the quantum behavior of electromagnetic radiation. Theorists reconciled the new quantum theory with Rutherford's newly observed subatomic structures and began the work that culminated in the standard model of particle physics. At about the same time, Einstein showed that the reconciliation of Maxwell's equations with Newtonian dynamics required a complete paradigm shift in our ideas about space and time that culminated in the general theory of relativity.

Physicists now faced the problem of reconciling the standard model with general relativity, and here computers have taken on a new and previously unexpected role. Not only are computers essential for operating the accelerators and other instruments that provide the data that the progress of physics ultimately depends on, as well for collecting, selecting, and analyzing the resulting data, but also computers have become essential for the development of theory itself. Modern theories have become too complicated to be developed with old-fashioned techniques from the birch-bark canoe era, that is, with pencil, paper, and chalkboard. There are several reasons for this, not the least being the fact that physicists over the centuries have to some extent exhausted the possibilities inherent in simple theories. Computers were already essential in the development of the standard model itself. The unification of the weak and the electromagnetic forces (as part of the standard model) depended critically on the detailed and extensive computer calculations of a property called renormalization. Those extensive calculations won a Nobel Prize for Veltman and t'Hooft in 2000. And Greene notes that "Kinoshita . . . has, over the last 30 years, painstakingly used quantum electrodynamics to calculate certain detailed properties of electrons. Kinoshita's calculations . . . have ultimately required the most powerful computers in the world to complete" (Greene, 1998, pp. 121–122). He goes on to describe similar calculations of his own that required the raw calculating power of computers to make similar involved calculations of particle masses in more recent theories (Greene, 1998, p. 276). And Frank Wilczek noted (2002, p. 125):

> There has been enormous progress in using QCD [Quantum Chromodynamics, a central component of what we have been calling the standard model] to describe the properties of protons, neutrons and other strongly-interacting particles. This involves very demanding numerical work, using the most powerful computers, but the results are worth it. One highlight is that we can calculate from first principles, with no important free parameters, the masses of protons and neutrons.

There is a deep mathematical reason why computer power is required for the development of modern mathematical physics. Modern theories tend to be nonlinear, and nonlinear theories are inherently complicated. Physicists have known for decades that although quantum mechanics is linear, any theory that incorporates or is merely consistent with general relativity must be strongly nonlinear, and its theoretical development is therefore likely to require large quantities of computer power.

It is not easy to grasp the difference between linear and nonlinear models without some detailed understanding of the mathematics involved. As I noted in chapter 1, the word "linear" simply means that there is some sense in which, if you plot or graph the behavior of the system while you vary some key parameter that controls the system, the resulting graph will be a straight line. But this fails to convey the vast gulf that separates linear problems from even some of the simplest of nonlinear problems. It is not too much of an exaggeration to say that linear problems are solvable, and nonlinear problems generally are not. Discussing systems of nonlinear algebraic equations, which is one of the simpler areas in which nonlinearities pose significant problems, Press et al. comment (1992, p. 372):

> We make an extreme, but wholly defensible, statement: There are *no* good, general methods for solving systems of more than one nonlinear [algebraic] equation. Furthermore, it is not hard to see why (very likely) there *never will be* any good general methods . . . in order to find . . . the solutions of our nonlinear equations, we will (in general) have to do neither more nor less than map out the full zero contours of both functions. . . . root finding becomes virtually impossible without insight! You will almost always have to use additional information, specific to your particular problem, to answer such basic questions as, "Do I expect a unique solution?" and "Approximately where?"

There is a another reason why nonlinear problems are vastly more difficult than linear problems, as described in Robertson (1998, pp. 6–50). Basically, the solutions to linear problems can be expressed with finite amounts of information, and the solutions to nonlinear problems generally cannot (except in cases with unusual symmetries). But although the full solution to nonlinear problems cannot be expressed with finite amounts of information, it is frequently possible to find special cases, particular domains that have solutions that are of interest that can be explored with computer power. And as Press et al. suggest above, such exploration generally involves massive amounts of computation to map out the region of interest.

Prior to the computer revolution, theorists were restricted by their limited calculating capacity and were generally confined to developing linear theories. They also developed techniques such as perturbation theories which attempt to linearize the calculations of inherently nonlinear phenomena by looking only at very short or small intervals over which the nonlinear behavior could be ignored. In contrast, computers allow us to do the calculations needed to handle nonlinear theories in a more direct fashion. It should be said that even with computers our ability to handle nonlinear theories is fairly limited, even though it is vastly greater than our ability to handle such theories without computers. The general study of nonlinear systems has spawned an entirely new area of mathematical research sometimes called "chaos theory." This ability to use computer power to explore and "see" the consequences of nonlinear theories has produced not only a new phase change but also perhaps a new kind of phase change—a phase change in which the novel ability to see things involves seeing things in a mathematical realm rather than in the physical realm. Computer power is proving essential to exploring the unavoidable nonlinearities of the theory(-ies) that is attempting to unify the standard model with general relativity.

The new theory that represents the latest and perhaps most promising effort to unify the standard model of quantum mechanics with general relativity goes by a number of different names. It was originally called string theory, then superstring theory, then M-theory. There is not space enough in the discussion here for a thorough development of this theory even if it were appropriate to do so. An excellent discussion written for general audiences can be found in Greene (1998). What we can do here is summarize some of the major paradigm shifts that are being generated by the present versions of M-theory, which is far from being completely understood.

Several of these paradigm shifts require substantial new adjustments to our concepts of space and time. If you thought that relativity theory was counterintuitive, you may be in for a rude shock: The new theory is considerably more complex and far less intuitive. Not only does it require eleven dimensions of space-time, as we will discuss below, but there is also a major change at the level of the very smallest structures of space and time. Theorists as far back as Aristotle had, in contrast, concerned themselves with questions about the largest scales of space-time, and Aristotle himself had speculated that there might be a maximum size to physical space, that the universe might be of finite extent, that is, it might have an outer edge out there somewhere. Einstein was the first to suggest that since space-time can be curved, it is possible that the universe is finite in size but without any outer edge anywhere, a three-dimensional analogue to the surface of a sphere which is a finite two-dimensional surface that also lacks an edge anywhere within the two-dimensional surface. But although researchers and philosophers had often theorized on the largest possible size in the universe, and others as far back as Leucippus and Democritus had speculated on the existence of the smallest possible particle of matter (the early speculations about the existence of atoms), no researcher in the prequantum era appears to have foreseen the idea that there might be a minimum size for space-time itself. Indeed, the classical concepts of Euclidean geometry forbid such a thing. Euclidean points have neither length, breadth, nor depth, and Euclidean geometry was held in nearly sacred awe until well into the nineteenth century.

But in M-theory there is such a thing as the smallest part of space-time, and this is an idea that may be more counterintuitive even than the idea of eleven dimensions. M-theory has a fundamental length scale called the Planck length, about 10^{-33} centimeters, and in this theory space simply does not exist on smaller scales. There is a corresponding smallest possible element of time called the Planck time. The fundamental Planck length can be calculated from a combination of the fundamental constants of physics. If you are mathematically inclined, you can calculate its value simply as

$$\text{Planck length} = \sqrt{\frac{hG}{2\pi c^3}}$$

where h is Planck's constant, G is Newton's gravitational constant, and c is the speed of light. And the Planck time is then simply the time required for a beam of light to travel the distance of the Planck length. The existence of a minimum scale for space-time removes many if not all of the difficulties encountered in the attempt to reconcile quantum mechanics with relativity, much as the transformation of space into time and vice versa removed the difficulties of reconciling Maxwell's electromagnetism with Newtonian dynamics.

As mentioned above, another major paradigm shift to come out of the current state of the development of M-theory involves the fact that space-time must have 11 dimensions if gravity is to be incorporated into a unified theory of all the known forces of nature. The details of the reasons for 11 dimensions cannot be easily compressed into a few paragraphs. Greene (1998) can be consulted for a more detailed account. Once you get used to the idea that the universe might have more than four dimensions then 11 does not seem quite so strange. But there remains the mildly uncomfortable fact that we normally perceive only three dimensions of space and one of time. The seven additional dimensions that M-theory requires must therefore have different properties than the ones we are used to, and again this should not seem too strange. We are already used to the idea that the three dimensions of ordinary space can have different properties.

For example, only two of the three dimensions that define an ordinary rectangular room are essentially identical, and the third is different. To understand what this means, suppose you are standing in one corner of an ordinary room holding a piece of chalk and someone asks you to walk across the room to the opposite corner and make a chalk mark on the floor. You would have little difficulty in complying. But if you were then asked to walk to the ceiling and make a mark there, you would have somewhat more difficulty. The vertical direction is different because of the presence of a physical field, a gravity field, that makes travel in the vertical direction more difficult than in the two horizontal directions. Similarly, your motion in the timelike dimension is even more constrained because of the presence of some physical effects (poorly understood at present) that force us to move in only one direction in that dimension.

What is the nature of these physical effects or forces that constrain our motion in time direction? Einstein's general relativity may provide a clue. According to general relativity, if you happen to be unfortunate enough to fall inside a black hole, you would find that the radial distance from the center of the black hole, which seems naively to be a purely spacelike dimension, is instead a timelike dimension. In other words, inside a black hole you cannot move upward. No matter what else you do, you must move steadily downward toward the center of the black hole. You have as little control over the rate of your motion in the downward direction (toward the center of the black hole) as the rest of us have control over our own motion in the timelike direction (see the discussion in Misner et al., 1973, p. 823 for the technical details). (Some theorists have speculated that our whole universe may itself be entirely inside a black hole of immense size. This speculation would take us a bit afield of the discussion here.)

Therefore, the idea that dimensions can be mathematically identical but physically different is already fairly familiar, although Einstein's original suggestion that timelike dimensions can be converted by transformations (similar to rotations) into spacelike dimensions and vice versa was uniformly regarded as madness when it was first proposed. And the nature of the forces that make the other seven spacelike dimensions of M-theory physically different from the four that we commonly perceive are not at all understood at present. The most popular suggestion is that they are in some way "curled up" so that they are extremely small in overall extent. Here the words "extremely small" refer to the Planck length, which is some 25 orders of magnitude smaller than atomic scales. Other theorists suggest that the extra dimensions might be large but that some presently unknown forces are preventing us from seeing them or traveling in those directions (Arkani-Hamed et al., 2002). Greene even raises the question of what might happen if one or more of the extra dimensions turns out to be timelike rather than spacelike (i.e., what happens if two or more dimensions are time-like?) (1998, pp. 204–205). No one yet understands what this might mean. In fact, it is safe to say that the M-theory is not well understood at present, in large part because of its extraordinary mathematical complexity. In addition to the nonlinearities of the theory, the mere existence of additional dimensions complicates the theory immensely, as does the existence of a minimum length scale. Our current formulations of M-theory will certainly have to be revised over the next few years to decades as theorists develop their ideas about the dimensionality of space and other aspects of the theory. And as we have seen, computer technology is playing an increasingly vital role in expanding our understanding of M-theory.

These discussions have covered only a very small part of the paradigm shifts caused by the phase change that has been generated by the use of computers in physics. Computers are essential to operating modern colliding accelerators, and the beam energy available with computer technology is six or seven orders of magnitude greater than it was without computer technology. And computers are also essential to modern detector technology, where computerized detectors can handle about nine orders of magnitude more information than was possible in the precomputer era. Finally, the unification of quantum mechanics with general relativity in M-theory is producing radical new paradigm shifts in our understanding of the universe and the nature of space and time, and computer technology has become indispensable for the development of the nonlinear mathematics needed to handle the inherent complexities of M-theory. It seems reasonable to say that progress in physics today would slow to a crawl if not grind to an absolute halt in the absence of computer technology.

But instead, computer technology today is producing the largest phase change in the history of physics by vastly expanding the range of phenomena that we can see, as well as by expanding the range of nonlinear models whose behavior we can see and experiment with. The comparable phase changes in the 1800s and early 1900s produced paradigm shifts related to quantum mechanics and relativity that have shaped the modern world and its concept of the universe. The newest phase change caused by the computer revolution should provide even more important and more radical paradigm shifts, related in part to a new understanding of the dimensionality of the universe and the small-scale structure of space-time. These paradigm shifts will probably extend our understanding in directions (both spatial and conceptual) that we cannot conceive of today. Indeed, Greene speculates that it may be our inability to formulate mathematical models that do not assume the preexistence of some form of space-time that may be the principal obstacle to progress in M-theory today (1998, pp. 376–380). The present era may well be the most exciting time in all history to observe the development of new concepts, ideas, and techniques in physics.

5

PHASE CHANGES IN MATHEMATICS

Mathematics may be the most difficult field in which to explore the phenomenon of phase changes, partly because the difference between a phase change and a paradigm shift is less clear here than in any other field. In fact, as Lax (1989, p. 539) noted, the computer is the only invention in the history of mathematics that closely parallels the invention of the telescope in astronomy and the microscope in biology, so it could be argued that there has been only a single phase change in the entire history of mathematics. The computer revolution fits both parts of the definition of a phase change in mathematics: it allows researchers to see things that could not be seen without it, and any attempt to extrapolate the characteristics of mathematics from the birch-bark canoe era to the computer age is destined to fail.

But if we take the position that the computer is generating the first and only phase change in the field of mathematics, then we immediately run into the difficulty that there is no other phase change to compare it against. For that reason I am going to compare the computer revolution to two earlier revolutions in mathematics that were generated by the works of Euclid and Newton, respectively, and ignore for the time being the somewhat blurred question of whether these earlier revolutions constitute phase changes or paradigm shifts. These two earlier revolutions clearly fit one part of the definition of a phase change, that no reasonable extrapolation of the field prior to the revolution comes close to describing the field afterward. And we can fit the other part of the definition, the ability to see something that could not be seen before, if we adopt a Platonist viewpoint. The basic idea behind Platonism is that mathematical facts are things that exist external to the constructions of human mathematicians. These external facts can be "seen" by developing novel and effective mathematical techniques. This is not a terribly controversial idea. As Bourguignon (2001, p. 176) noted: "the vast majority of mathematicians are closer to the Platonist viewpoint than to the other one, or at least '*they spend a good portion of their professional time behaving as if they were,*' as Andre Weil once put it."

A significant part of the difficulty in exploring the phase change associated with Euclid's work is that the historical records of the development of mathematics prior to Euclid's time are fragmentary, almost nonexistent. Thus, any attempt to extrapolate any earlier work in mathematics is somewhat problematic. It is fairly clear that much of the early development of geometry took place in Mesopotamia and especially Egypt, but there are almost no surviving records of that development.

In Egypt the annual flood of the Nile certainly created some unique problems for marking boundaries that may have driven the early development of geometry. In other parts of the world, physical objects such as stones and trees could be used to mark boundaries, as Euripides notes in *The Trojan Women,* where Cassandra muses on the unusual nature of the cause of the Trojan War with the comment: "No man had moved their landmarks" (quoted in Hamilton, 1965, p. 47). But the use of boundary stones and other physical landmarks ran into serious difficulties in Egypt, where the flood of the Nile would render such markers ineffective by burying them in silt or even washing them away. The need to reestablish boundaries after each repeated flood apparently led the Egyptians to develop geometric techniques for surveying and measuring the land. The very word "geometry" preserves the root meaning of measuring the Earth. Their practical techniques for surveying the land included such things as the use of ropes measured in 3–4–5 unit ratios to construct accurate right triangles. This might be considered a phase change in the sense of "seeing" some of the elementary facts of geometry and right triangles. However, it is fairly clear that although the Egyptians developed a lot of practical knowledge of geometry, they failed to develop any rigorous or coherent body of geometric theory.

At one stroke Euclid changed everything. Although earlier mathematicians including Hippocrates and Thales had employed early forms of proof using logical arguments, Euclid was apparently the first to realize that everything in geometry could be derived using logical operations on a small number of axioms. This defined the entire field of geometry and indeed the rest of mathematics in a way that would remain recognizable after more than two millennia, right down to the present day. In no other field of science are techniques that were developed in the first millennium B.C.E. still employed routinely today. As Dunham put it (1990, p. 32):

> To this day . . . mathematicians will first present the axioms and then proceed, step-by-step, to build up their wonderful theories. It is the echo of Euclid, 23 centuries after he lived.

However, as in many other disciplines, the methods pioneered by Euclid ran into some serious difficulties and major limitations in the precomputer era. Today the complexity of the problems that mathematicians are trying to solve has begun to outstrip the time and resources that are available for calculations, proof-construction, and proof-checking techniques by hand, the methods that Euclid would have recognized. One of the first clues to the magnitude of the problem came with the celebrated four-color theorem. The genesis of this problem dates to the middle of the nineteenth century. As I noted in an earlier book (Robertson, 1998, pp. 65–66):

> A graduate student named Francis Guthrie apparently stumbled on this problem when he asked whether four colors are sufficient to color any map. In other words, he asked whether it is possible to divide up a Euclidean plane into areas that could not be colored with four colors, such that no areas having a common border have the same color. Mathematicians as distinguished as De Morgan and Hamilton tried to prove this theorem without success. The problem became infamous as "The most easily *stated,* but most difficult to answer, open question in mathematics" (Stewart, 1992, 105). Oddly, the problem was solved first for complicated surfaces such as a torus (doughnut-shaped surface) where seven colors were found to suffice. The proof of the four-color theorem on the plane was finally attained by Haken and Appel in 1976 using 1200 hours of time on a then-powerful computer. The novel thing about Haken and Appel's proof is that there is no way for a human mathematician to check the computer's work. It would take too many lifetimes.

Haken and Appel's proof involved an exhaustive and ultimately futile search for a counterexample. This method of exhaustive demonstration that no counterexample exists is one form of mathematical proof. But other more powerful techniques are needed in order to use computers to develop more general results in mathematics. The critical early breakthroughs along these lines were made by Alan Turing and Claude Shannon over half a century ago. Turing was the first to realize that all the operations of formal logic (including every operation that can be defined in mathematics) can be performed by a machine, and he carefully defined the minimum necessary characteristics of such a machine. Turing's work provides the fundamental theoretical underpinnings of the theory of the operation of electronic computers today. And Shannon showed that all the operations in digital electronic circuitry could be modeled as operations in formal logic and Boolean algebra. Therefore, all the operations that mathematicians carry out in the process of constructing a proof can in principle be carried out by modern computers. Computers can carry out these logical operations not only at speeds that are many orders of magnitude faster than human mathematicians but also with error rates that are minuscule if not zero, far smaller than the error rates that can be attained by even the most careful of human mathematicians.

As explained in some detail by Mackenzie (1995), the modern application of this vast new power to develop axiomatic methods of proof has proceeded in several directions. One line of research involves simulating human forms of deduction. The second involves developing new forms of logical argument that are difficult for human mathematicians but tailored to the capabilities of computers. This second approach centers on an algorithm called "resolution." The resolution algorithm has produced some stunning successes, including such things as finding a proof of a conjecture in Boolean algebra made by Herbert Robbins in the 1930s. For decades the proof of the Robbins Conjecture had eluded the efforts of some of the most celebrated mathematicians of the twentieth century, including legendary theorists such as Alfred Tarski. But in 1996, a computer search found three different proofs of the Robbins Conjecture using six hours of computer time, and each of the three proofs was simple enough to be checked by mathematicians by hand.

But despite these successes computerized proof techniques all ran headlong into a serious, seemingly intractable problem. Although it is possible in principle to find a proof by programming a computer to check all possible combinations of logical operations on axioms, the number of possible sequences of logical operations that the computer must explore grows explosively with the number of steps in the proof.

The problem of finding a proof can therefore very quickly exhaust even the vast computational power of modern computers. Suppose, for example, there are 10 possible logical operations at each step of the proof (there are generally more than this), and the entire proof requires only 30 steps, a very simple proof indeed. But to search out the consequences of every possible combination of logical steps would require exploring 10^{30} possible paths. This enormous number is roughly comparable to the number of protons that would span the visible universe. No computer could handle such an enormous number of possibilities.

This problem is sometimes referred to as a "combinatorial explosion," and a third general method for constructing computerized proofs has been developed to try to deal with it. This third method is referred to as "interactive theorem proving"; in this technique, the computer proof is directly guided by a human mathematician. In other words, instead of having the computer search all the 10^{30} possibilities in the proof above, the human researcher would try to find intermediate results that require smaller numbers of steps. By analogy, suppose a computer is trying to find an automobile route between Denver and Indianapolis, and it has access to a database showing all the roads in the United States. The computer could search all possible roads until it finds a combination that starts in Denver and ends in Indianapolis. Or it could be told by a human partner that the route should pass through Kansas City and then St. Louis. The computer then searches for routes from Denver to Kansas City, then Kansas City to St. Louis, and then St. Louis to Indianapolis. By breaking the trip into three legs, the number of combination of roads that need to be to be searched can be made much smaller.

Suppose, for example, in an interactive proof the human mathematician were able to devise two intermediate results that would break the 30-step proof into three separate 10-step segments (analogous to knowing that the car route above must pass through Kansas City and St. Louis). The computer would then need to search only 10^{10} possible routes for each segment, for a total of 3×10^{10} steps or 30,000,000,000 (thirty billion) steps instead of 1,000,000,000,000,000,000,000,000,000,000 steps for the entire proof, an enormous saving of computer time. Thirty billion steps is still beyond human capabilities, but it is not beyond the power of computers. Thus, these interactive proof techniques have the potential to bring the impossible down into the realm of the merely difficult.

Interactive computerized proofs may actually resemble conventional mathematical practice more closely than even most mathematicians realize. Nidditch (1957, p. v) states: "In the whole of mathematical literature there is not a single valid proof in the logical sense." Nidditch does not mean that most proofs in the literature are wrong, only that they are incomplete. Davis (1998, p. 171) refers to the practice of omitting steps in a proof as "skipping." Mathematicians generally focus, quite reasonably, on the critical steps of a proof, the steps that require special insight and novel ideas. They often neglect the routine intermediate steps that would be needed to make the proof fully rigorous. Laplace was notorious for leaving out sometimes lengthy intermediate derivations that were obvious to him but frequently less obvious to his readers. Modern interactive computer proof techniques will allow mathematicians to continue to focus even more precisely on the critical, difficult steps in a proof, but now they can have complete confidence that all of the routine intermediate steps will be worked out with full rigor and recorded by their computer assistants.

One of the huge advantages to both interactive and noninteractive computerized proof search techniques is that the final proof is immediately in a form that can be checked by computer. Indeed, if the computer software and hardware have functioned without error, then no check would be required. But because errors are possible in both hardware and software, the final proof will generally have to be checked with independent versions of both hardware and software. It will never be possible to reduce the chance of error completely to zero, but it will be straightforward to reduce the probability of error to a value that is many orders of magnitude smaller than the probability of error for a human mathematician checking the same proof. Indeed, the time and effort that mathematicians once spent worrying about errors and checking for them will be largely eliminated with computerized proof techniques. This will allow mathematicians to focus far more of their valuable time on the critical and genuinely difficult portions of a proof. Work is already underway to develop standardized, canonical forms for transmitting mathematical information between computers and across the internet in ways that can be understood and checked by a variety of different software packages (A. M. Cohen, 2001).

Thus, computers are causing a phase change in mathematics similar to the one produced by Euclid with the introduction of axiomatic methods. But how can we measure the magnitude of these two phase changes? If we are going to argue that the computer revolution is more important than the Euclidean revolution, then we ought to be able to show that the magnitude of this phase change is significantly larger than the phase change that Euclid engendered. It should be admitted that quantitatively measuring the ease of producing a proof is not as easy as measuring the angular resolution of a telescope, for example. But we can cast the problem into a simple numerical question: How many proofs can be attained with computer techniques, and how does that number compare with the number that would be attained with conventional hand-labor techniques? In reducing this question to a simple number I am assuming that all proofs are equally interesting and important, an assumption that is clearly false. Yet it seems reasonable to assume that "interest" and "importance," assuming they can be defined, are roughly constant when averaged over very large sets of proofs.

This question of the magnitude of the two phase changes has two distinct answers, one related to the checking of extant proofs and the other related to the generation of those proofs in the first place. The effort involved in checking proofs is roughly linear in the number of steps in the proof, because each step in the proof has only two possible results, correct or incorrect. If any step registers as incorrect, then the proof check is finished (although at that point the proof might be modified to fix the problem). The difference between the size of proofs that can be checked by computers and the size that can be checked with hand methods should therefore scale directly with computer speed. For example, if a computer is a billion times faster than a human, it should be able to check a proof that is about a billion times longer than the one that the human mathematician can check in the same amount of time.

It may seem strange that mathematicians have been reluctant to embrace computer proof-checking techniques since, as Tymoczko (1998, p. 248) explains: "Genius in mathematics lies in the discovery of new proofs, not in the verification of old ones." Indeed, it would seem that checking proofs is an ideal application of the vast calculating power of computers. Checking proofs by hand is hard, tedious, and error-prone labor. And for proofs that involve a large number of steps, the probability that a human mathematician will make a mistake is vastly greater than the probability that a computer will make a mistake. Thus, we can expect that a time will come in the not-too-distant future when proofs will not be generally accepted within the mathematical community until they have been verified by a number of separate proof-checking programs. Yet as Mackenzie (1995, p. 21) notes, to date "interactive theorem provers and automated proof checkers have generally not taken root within mathematics itself," although these programs have had broad applications in verifying computer programs and hardware designs. These verifications share many of the characteristics of mathematical proofs. Of course, mathematicians have two basic objectives in checking a proof. The first is simply to satisfy themselves that it is correct. The second is to learn how the proof works, what are the critical steps and ideas. The first of these objectives is completely attained by computer verification. And in the case of interactive proofs, a good fraction of the second objective may also be attained.

Although checking a proof is a linear problem, finding the proof in the first place is a problem whose difficulties scale exponentially with the length of the proof because, as noted above, the number of alternatives that need to be searched grows by a roughly constant multiplicative factor with each step. The ability to find proofs should therefore scale as the logarithm of the computer speed. This is a huge penalty. For computer speeds in the range of a billion times faster than human calculations, it implies that computers could find proofs that are in the range of only about ten times longer than what humans are able to do. (It is clear from this calculation that computerized proof techniques will be limited by the difficulty of constructing proofs, not by the difficulty of checking them.) However, the number of steps in a proof is perhaps not the best measure of the power of computers—there is another exponential function to consider because there are a lot more long proofs than there are short proofs. Thus, if computers are limited to finding proofs that are only about ten times longer than can be found without them, this is far more than just ten times more proofs. Furthermore, with interactive proof techniques the search space can be reduced substantially, and the entire logarithmic penalty need not be paid in every case.

Because there is no limit in principle to the number of axioms needed for a proof, nor is there any maximum limit on the number of logical steps in a proof, it is clear that nearly all the proofs that are possible in mathematics are too complicated for human mathematicians and also too complicated for the fastest possible computers. But even though nearly all proofs are forever out of reach, still the number of proofs that can be found and checked only by using powerful computer techniques is vastly larger than the number that can be found and checked "by hand," simply because computer operating speeds are many orders of magnitude faster than human speed. It is therefore reasonable to think that the ratio of the magnitude of the phase change introduced into mathematics by Euclid to the magnitude of the similar phase change introduced by computer technology will be something on the order of the ratio of human to computer calculating speeds, within some orders of magnitude. It would be fairly pointless to try to pin down an exact numerical value for this ratio because the value is going to change rapidly with the development of new computer technology and proof algorithms. Still, it cannot be seriously doubted that the volume of "proof space" that can be explored with computers is many orders of magnitude greater than can be explored without them. Yet many mathematicians have expressed a great deal of resistance to this idea. The following reactions may not be universal, but they appear to be representative. As Langtangen and Tveito put it (2001, p. 813):

> pure mathematicians in general seem to ignore or sometimes fight against the possibilities offered by computers.

Mackenzie (1995, p. 24) quotes D. I. A. Cohen criticizing the four-color proof as being

> "computer shenanigans." Their "proof" did not "explain" as any real proof must: "it didn't tell you *why* four [colors] is the answer. Mathematics is supposed to do exactly that: give you understanding."

Mackenzie (1995, p. 24) quotes Bonsall as arguing against computerized proofs as follows:

> We cannot possibly achieve what I regard as the essential element of a proof—our own personal understanding—if part of the argument is hidden away in a box. Let us avoid wasting . . . funds on pseudo mathematics with computers and use them instead to support a few real live mathematicians.

The crux of the problem was perhaps best stated in a quote attributed to Eugene Wigner (Nieminen, 2001, p. 938):

> It is nice to know that the computer understands the problem. But I would like to understand it, too.

In short, mathematicians generally feel that computer proofs must be surveyable, that is, they must be comprehended in toto by human mathematicians. But this seemingly innocuous requirement leads to serious problems because it deliberately ignores the vast realm of proofs that can be explored by computers but not by the "naked" human mind. The attitude of these mathematicians is closely analogous to that of many scholars in Galileo's day who refused to look through his telescope. They wished to base astronomy solely on observations made with the naked eye, as astronomers had always done in the past. A recent biography describes Galileo's response to this line of argument (Geymonat, 1965, p. 45):

> Besides objections based on the poor functioning of the lenses then in use, Galileo had also to overcome some others of a completely different sort. These, which were certainly no less dangerous, were linked to the belief (shared by most scholars of the time) that only direct vision had the power to grasp actual reality. . . . To overcome this latter type of objection required philosophical arguments to expose the absurdity of making our eyes the absolute criterion of real existence. Galileo wrote: "Besides, who would wish to say that the light of the Medicean planets [Galileo's name for what are now termed the Galilean satellites of Jupiter] does not arrive on the Earth? Are we to make our eyes the measure of the expansion of all lights, so that wherever the images of luminous objects do not make themselves sensible to us, it should be affirmed that light does not arrive from them? Perhaps such stars, that remain hidden from our weak vision, are seen by eagles or lynxes." . . . Today, doubts of this kind may seem infantile; to appreciate Galileo's achievement, it is necessary to take into account that they constituted extremely grave difficulties in his time.

Just as astronomers had to accept the idea that the telescope vastly extends the reach of the naked eye, mathematicians will have to accept the idea that the computer similarly extends the reach of the naked human mind. It is not really wrong to refuse to use computerized search and verification techniques for mathematical proofs, it is merely less than optimal, less than the best use of our available resources. And mathematicians who refuse to learn and utilize computerized techniques can expect to be at a serious disadvantage compared to those who do use such techniques, just as astronomers who refused to use telescopic observation techniques were at such a disadvantage that they quickly became essentially extinct. After all, as I noted (in Robertson, 1998, p. 68): "mathematicians are just like the rest of us: They lust after power. And computer techniques represent a level of power that they will not be able to resist for long." The region of "proof space" that lies beyond the reach of the naked mind, that can be explored only with computer techniques, may well contain proofs that are simply too intrinsically interesting to ignore, just as those reaches of the astronomical universe that lie beyond the capabilities of the naked eye contain objects that are too intrinsically interesting to ignore.

As noted above, there are two important aspects of a mathematical proof: First, a proof tells us that certain theorems are correctly deducible from certain sets of axioms; second, the proof helps us to understand the reasons why those theorems are deducible. Both of these aspects are important, but both of them may not always be attainable. For complex computer-generated proofs, only the first aspect may be attainable. We should not throw this information away merely because the second aspect is beyond our reach. And finally, we should note that there will be cases in which, although the complete computer proof lies beyond our ability to understand it, we can still understand many or all of the critical steps, and we can understand the computer code that is used to generate the vast number of steps that fill the gaps between critical points. The proof of the four-color theorem is in this category. Therefore, the fact that the totality of a proof is too complex for us to work through in one lifetime does not by itself imply that we cannot attain a substantial understanding of at least the critical elements of the proof.

The use of computers to explore the vast unknown realms of mathematics that lie beyond the reach of the unaided human mind should provide not only the largest phase change in the history of mathematics, but it may also well be the most important phase change in all of science, if for no other reason than the fact that mathematics has broad applications in all the other sciences. It would be ironic and more than a little amusing if, owing to their resistance to the use of computer techniques, mathematicians turned out to be the last ones to realize this.

I noted at the beginning of this chapter that there is a second revolution in the history of mathematics that might be thought of as a phase change, the revolution that followed Newton and Leibniz's invention of calculus. This revolution has also led us recently to sets of problems that have a level of complexity that simply cannot be handled with hand-labor techniques. Calculus was itself a set of powerful new techniques that were devised to deal with some problems that had troubled mathematicians from antiquity. Even before Euclid's time mathematicians had known how to calculate the area of a plane figure whose sides consist entirely of straight lines. Calculating the areas of square and rectangular figures was easy, triangular figures were only slightly more difficult, and any figure that was completely bounded by straight lines could be cut up into triangular pieces and the areas of the resulting triangles could be calculated and added up. But, of course, many interesting geometric figures are not bounded by straight lines, most obviously the circle. The circle was of central interest to early Greek geometers, and not being able to calculate its area was something of an embarrassment for them.

The first known case of the calculation of the area of a figure with curved sides occurred several generations before Euclid. Hippocrates managed to calculate the area of a crescent-shaped figure that was bounded by two circular arcs, but he used methods that could not be easily generalized to other curved figures. It was Archimedes of Syracuse, the greatest mathematician of antiquity, who built on the earlier work of Eudoxus to calculate both the area of the circle and the area under a parabola and then went on to calculate the volume of a sphere and the area of the surface of a sphere (see Dunham, 1990, chapters 1 and 4). But following Archimedes no further progress was made for nearly two thousand years.

In the late 1600s Newton and Leibniz simultaneously developed calculus, the first set of general techniques for calculating the area of figures that were bounded by curved lines, a procedure that is referred to as "integration." Calculus also dealt with the related problem of determining a tangent line to an arbitrary curve, a straight line that touches the curve at only a single point, a procedure that is referred to as "differentiation." Newton and Leibniz discovered, to almost everyone's surprise, that the problems of differentiation and integration are inverses of each other. And Newton realized that these new mathematical techniques had extraordinary power far beyond the straightforward matter of calculation of areas and tangents. He used calculus to develop an entire theory of dynamics which, coupled with his inverse-square law for gravity, he was able to use to derive Kepler's results about planetary motions and also explain a string of other phenomena, including ocean tides and the precession of the equinoxes.

Fundamentally, the reason that calculus is so extraordinarily important in mathematics is that calculus made it easy to solve problems that had previously been difficult and made it possible to solve problems that had previously been impossible. One obvious example of such a problem is the calculation of the area of a circle; this problem had been so difficult that its solution required the greatest mathematical genius of antiquity; calculus reduced the problem to an elementary exercise for high-school students. And problems such as the orbital motions of the planets that had not been adequately solved by any mathematician prior to Newton, have been converted into exercises for undergraduate physics students.

The phase change that is associated with the computer revolution has this same property but on a different scale: With computer techniques problems that had previously been difficult became easy, and problems that had been impossible became solvable. Even inside the domain of calculus itself, the range of problems that can be solved with computers is vastly larger than the range that could be handled without it. As Bailey and Borwein (2001, p. 52) observed, commercial software packages that perform operations in symbolic algebra and calculus "routinely and correctly dispatch many operations that are well beyond the level that a human could perform with reasonable effort."

In fact, such packages employ a newly developed integration algorithm called the Risch algorithm. As Bronstein (1997 p. vii) puts it:

> Yet, while symbolic differentiation is a rather simple mechanical process, suitable as an exercise in a first course in analysis or computer programming, the inverse problem has been challenging scientists since the time of Leibniz and Newton, and is still a challenge for mathematicians and computer scientists today. . . . most calculus and analysis textbooks give students the impression that integration is at best a mixture of art and science, . . . The goal of this book is to show that computing symbolic antiderivatives [integrals] is in fact an algorithmic process.

The Risch algorithm renders conventional calculus techniques, such as integration by parts and substitution techniques (which are still taught in many elementary calculus courses), as obsolete for use in calculus as Roman numeral techniques are for problems in long division. As Lax (1989, pp. 539–540) noted: "dissatisfaction with the traditional calculus is nearly universal today. . . . This welcome crisis was brought on by the widespread availability of powerful pocket calculators that can integrate functions, . . . and solve differential equations with the greatest of ease, exposing the foolishness of devoting the bulk of the calculus course to antiquated techniques that perform these tasks much more poorly or not at all." When the Risch algorithm was first implemented on computers in the latter part of the twentieth century the first thing its developers did was to check it against standard tables of integrals. To their amusement and horror they found errors in the tables; one standard set of tables had an error rate of about 25% (Pavelle et al., 1981, pp. 136–153).

Just as in the case of computerized proofs, it would be pointless to try to get an exact measure of the size of the phase change associated with computerized symbolic algebra and calculus techniques. The magnitude changes constantly as new hardware and software are developed. But there can be little doubt that this phase change is many orders of magnitude larger than the phase change associated with the original invention of calculus. In other words, the number of problems even inside calculus itself that can be solved with modern computer techniques is vastly larger than the number that can be solved in calculus without computer technology.

But there is yet another dimension to the phase change in mathematics engendered by the computer revolution. If computers have vastly expanded the set of problems that can be solved within calculus itself, they have also had enormous impact in other areas of mathematics. This is yet another justification for asserting that the phase change in mathematics associated with computer technology is much larger than the corresponding phase change that was associated with the invention of calculus. Mathematicians have given the name "analysis" to the general class of problems that calculus deals with. Roughly speaking, analysis deals with functions or curves that are continuous and smooth. But there are other broad areas of mathematics, including such things as the theory of numbers, finite mathematics, and combinatorics that calculus touches only indirectly, if at all. Computer techniques are already making significant contributions to many of these areas and have the potential to revolutionize all of them. This is a straightforward result of the extraordinary power of Alan Turing's insight that every operation that can be defined in mathematics can be performed by a machine.

One of the ways in which raw computer power has generated major changes in mathematics lies in the stunning power that computer graphics has for generating insights and driving important discoveries. As Bailey and Borwein (2001, p. 53) put it: "We would be remiss not to mention the growing power of visualization especially when married to high performance computation." Perhaps the most familiar use of computer graphics lies in exploring the behavior of fractals, especially the Mandelbrot set and Julia sets, whose pictures have become almost cliches on posters and book-covers. And Hitchen (2001,

p. 588) notes in a discussion of new developments in differential geometry: "the satisfaction of seeing these surfaces before our eyes using computer graphics is also a great motivation for advancing the analytical side of the theory."

The staggering calculating power available with computer technology has led to the development of an entirely new branch of mathematics sometimes called "experimental mathematics." As Bailey and Borwein (2001, pp. 51–52) state: "pure mathematics (and closely related areas such as theoretical physics) only recently began to capitalize on this new [computer] technology. . . only in the last decade . . . has this technology reached the level where the research mathematician can enjoy the same degree of intelligent assistance that has graced other technical fields for some time. . . . There is now a thriving journal of *Experimental Mathematics*."

Bailey and Borwein go on to describe a remarkable sequence of mathematical identities that have been discovered with novel algorithms for finding integer solutions to problems. In many cases, the process would begin with extensive high-precision calculations that would determine that a particular solution is approximately an integer to an accuracy of some dozens of decimal places. Then more refined integer relation calculations would be used to show that the result is, in fact, exactly an integer. Bailey and Borwein (2001, p. 57) note that "this illustrates neatly that one can only find a closed form if one knows where to look." Or at least it is very much easier to find the closed form when one knows exactly where to look.

One of the very interesting formulas that was discovered using months of computer time with these techniques is the following series approximation that converges on π with remarkable speed:

$$\pi = \sum_{k=0}^{\infty} \frac{1}{16^k} \left[\frac{4}{8k+1} - \frac{2}{8k+4} - \frac{1}{8k+5} - \frac{1}{8k+6} \right]$$

The convergence of this series is so fast that the first two terms alone provide an accuracy of five parts in 10^5. Bailey and Borwein (2001, p. 55) claim that "this is likely the first instance in history that a significant new formula for π was discovered by a computer." And this formula has been used to develop an even more remarkable algorithm that allows one to calculate any digit in π using low-precision arithmetic without calculating any of the other digits. This extraordinary algorithm has been used not only to calculate individual digits of π

out past the trillionth decimal place, but it can also be used to investigate some long-standing problems including the question of whether the digits in π are "normal," which means roughly that every n-digit string of digits occurs with the same frequency as every other n-digit string. An algorithm that can calculate any given digit provides an important new analytic tool that can be brought to bear on this problem.

In another remarkable case, an important combinatorial identity was discovered as a result of a typographical error in the input to a symbolic algebra program. Bailey and Borwein (2001, p. 57) describe this as "the computational equivalent of discovering penicillin after a mistake in a Petri dish." The researcher had inadvertently posed a much more difficult problem than the one that was actually intended and was startled to discover that the computer was able to solve the much harder problem.

Perhaps the most stunning and important impact of computer technology on mathematics has been its effect on the study of logic itself. It has been argued that the existence of electronic computing machines may force changes in our approaches to formal logic. MacKenzie wrote (1995, pp. 329–330):

> There is reason to think that formal logic, long shaped by the exigencies of its relationship to mathematics, is now starting to march to the beat of a different drummer: computer science. . . . Nonclassical logics, unsuitable for modeling mathematical reasoning, are no longer simply philosopher's playthings but are being used to model the operations of machines in simulations in which those involved often believe that the use of classical logic would be cumbersome or incorrect. . . . It may well be that the modeling of machines will require richer logics than the modeling of mathematics, and that logic itself will be changed fundamentally.

MacKenzie points out that there have always been different groups of logicians with radically differing ideas, and some computer scientists have found that the constructivist ideas of L. E. J. Brouwer, which were long opposed by Hilbert and his formalist allies, are more conducive to thinking about programming problems than is more conventional logic.

This shift in emphasis in formal logic is part of a more general change that is taking place in mathematics today, a shift that is perhaps comparable to the one that took place in the sixteenth to the eighteenth centuries when mathematicians such as Cardano, Descartes, Newton, and Leibniz began to shift away from Euclidean geometry and focus instead on the more abstract areas of algebra, calculus, and analysis. Today many mathematicians are shifting away from conventional topics such as analysis, modern algebra, topology, and group theory that mathematicians have been investigating for centuries and instead are investigating problems related to computation, including such things as computational efficiency and the nature and limits of algorithms. This new shift is being driven partly by the enormous advances in computational power provided by the computer revolution, and it is typified by the work of mathematicians such as Goedel, Turing, Von Neumann, and Chaitin. As Chaitin (2002, pp. 135–136) wrote:

> I think that computers are changing the way we do science completely, and mathematics too. The computer can provide an enormous amplification of our own mental abilities and it's really changing the way everything is done. . . . The computer changes the way you think about things. One way to say that is that you only understand something if you can program it. Another way to say it is that the computer is the empirical content of mathematics. The computer is the lab for mathematics, the same way that the physics lab is the empirical content for physics.

The overall importance of this new shift in direction for mathematics can be gauged by the broad level of interest that has been generated outside of mathematical circles by discoveries such as Goedel's incompleteness theorem.

These computerized discoveries in mathematics are exactly analogous to Galileo's use of a telescope to discover such things as the satellites of Jupiter and sunspots on the Sun. Galileo's discoveries were stunning in their own right, but they gave only a feeble clue to the vastness and the wonders of the universe that lie far beyond the reach of the naked human eye, wonders that would be revealed by telescopic observations over the next few centuries. Similarly, these computerized mathematical results have given us the very first dim glimpses into the vast realms of mathematics that lie far beyond the reach of the naked human mind, beyond the region that mathematicians of the birch-bark canoe era were capable of reaching but which are suddenly within our reach through the use of the computational power of computers. As Bailey and Borwein (2001, p. 63) state: "We are only now beginning to experience and comprehend the potential impact of computer mathematics tools on mathematical research. In ten more years, a new generation of computer-literate mathematicians, armed with significantly improved software on prodigiously powerful computer systems, are bound to make discoveries in mathematics that we can only dream of at the present time." We have good reason to expect that many dazzling, unexpected, and counterintuitive results will soon be found as mathematicians begin to explore this vast and wonderful new universe.

6

PHASE CHANGES IN THE EARTH SCIENCES

In the earth sciences we will find the same patterns that characterize all the other fields we have examined. "Seeing" inside the Earth is one of the hardest problems that scientists ever faced. But a sequence of critical inventions caused phase changes by allowing us to see and detect structures inside the Earth that had never been seen before. These critical inventions made it possible to determine the overall structure of the interior of the Earth. And modern computer technology today is vastly expanding our ability to see and detect the details of the structure of the Earth.

The study of the Earth is a subject whose origins stretch back into antiquity. Early cultures developed a number of ideas about the size and shape of the Earth, many of which seem bizarre and even amusing today. The Earth was frequently perceived as a flat object of varying shape poised on the backs of turtles and elephants and such things. Philosophers in classical Greece did rather better than other early cosmographers. By Aristotle's time, they had deduced from observations of the shadow of the Earth on the Moon during a lunar eclipse that the Earth had a spherical shape, and Eratosthenes made a famous measurement of the size of the Earth by observing the differences in the lengths of shadows cast by the Sun at two different latitudes.[1]

Therefore the Greek cosmographers had a fairly good understanding of the shape and size of the Earth, and for millennia that was essentially all that was known about the Earth, apart from an increasing knowledge of the geography of its surface. But, as we noted above, gathering additional information about the Earth and its interior structures turned out to be one of the most difficult problems that scientists have ever encountered, far more difficult than acquiring information about the astronomical universe, for example. The naked human eye can detect objects that are millions of light-years away (the Andromeda Nebula at a distance of three million light-years is not difficult to observe under a dark sky), and modern astronomical instruments can observe clear across the visible universe, all the way out to the cosmic blackbody background radiation, which is essentially the flash of radiation left over from the big bang itself, the start of the entire universe. But there was no obvious way to observe in the downward direction, to see what lies under our feet rather than over our heads. Light cannot penetrate more than a few millimeters into the Earth's crust, and mines, wells, and other holes that penetrated into the Earth seldom reached depths of even a few kilometers, a tiny fraction of the 6,400 kilometer distance to the center of the Earth. In principle, you could examine rocks that have been formed at depth, but the physical and chemical knowledge needed to analyze and interpret such rocks is a fairly recent development. Thus, there was no easy or obvious way to learn anything about the composition and physical properties of the Earth far below its surface. And even observations at the Earth's surface were able to cover only about a quarter of that surface. Three-quarters of the Earth is covered by oceans, and it was not realized for centuries that the portion of the Earth's surface that forms the floor of the oceans is radically different from the portion that is seen on the continents, and furthermore that the ocean floor held many of the keys to understanding the overall structure of the globe.

Indeed, the ocean floor was uniformly regarded as uninteresting by geologists up to the middle of the twentieth century. It was generally thought to be an abyssal plain that was devoid of interesting features because it was covered in sediment that had accumulated gradually over billions of years. As one geologist commented later, it is amazing how simple the ocean floor seemed to be before we knew anything about it. Not until late in the nineteenth century did geophysicists begin to develop effective techniques for observing and measuring both the inside of the Earth and the Earth's surface under the oceans. The phase changes in geophysics that we will examine here will center on the ability to see things in the Earth and under the oceans that could not be seen prior to each phase change.

In spite of the difficulties of observing below the surface of Earth, scholars have been interested in the internal structure and functioning of Earth for millennia. Much of this interest was naturally centered on the phenomenon of earthquakes; the massive destruction that is caused by an earthquake is difficult to ignore. Two of the regions in which civilization originally emerged and for which we therefore have detailed records from antiquity, the eastern Mediterranean region and east Asia, are both located in earthquake-prone regions of the globe. Early ideas about earthquakes were summarized by Agnew (2002). Not surprisingly, earthquakes were commonly viewed as having supernatural causes. In Greek mythology Poseidon was referred to as the "Earth-shaker." China kept particularly detailed records of earthquakes because natural disasters were considered to be important evidence of celestial disfavor with the current dynasty. In 132 A.D., Zhang Heng invented the first seismoscope, a device for detecting earthquakes. Aristotle tried to devise a physical explanation for earthquakes in terms of the effects of meteorological conditions and winds in the Earth's interior. In more modern times, the great earthquake of 1755, which killed an estimated 60,000 in Lisbon, Portugal, stimulated interest in earthquakes among European scientists and led to much speculation about earthquakes being caused by chemical explosives or steam explosions. This interest led over the next century and a half to the development of crude seismometers to measure ground motions. Because of their low sensitivity, these early seismometers could only measure the major shaking of the ground in the vicinity of earthquakes, what today would be referred to as "strong motion" of the ground.

But it was not generally realized that earthquakes held the key to seeing the internal structure of the Earth, and this newfound ability to "see" the physical properties of the interior of the Earth marks one of the major phase changes in the history of the earth sciences. The story begins with a critical breakthrough that was made by Ernst von Rebeur-Paschwitz in Potsdam, Germany, in 1889. As is the case with many critical breakthroughs, this one was made in the course of an observing program that was designed to investigate something else entirely. The principal research interests of von Rebeur-Paschwitz lay in astronomy, not in the study of Earth. He was investigating tides, the motions and deformations of the oceans and (less familiarly) of the solid Earth caused by the gravity fields of the Moon and the Sun. Von Rebeur-Paschwitz had built a very sensitive pendulum to measure the rather small effects of these tidal forces. His pendulum would shift very slightly in response to changes in the direction of the gravitational attraction of the Sun and the Moon as they move relative to the Earth underneath them.

On April 18, 1889, von Rebeur-Paschwitz noticed that his pendulum showed a transient disturbance, a shaking motion, as though something had shaken the laboratory itself. When he later learned that a major earthquake had been reported in Japan a few hours earlier he immediately recognized the significance of the fact that his laboratory had been shaken by an earthquake that was thousands of kilometers distant, as well as the significance of the fact that such shaking could be observed with sufficiently sensitive instruments. The vibrations from the earthquake had traveled through the Earth itself. Since these vibrations are a form of radiation (radiating energy) that penetrates into and throughout the entire Earth, they could be used to probe and measure the internal properties of the Earth itself, much as X rays are used to observe structures inside the human body or Rutherford's alpha particle radiation experiments were used to probe the internal structure of the atom (see the discussion in chapter 4).

Von Rebeur-Paschwitz was the first to identify a distant earthquake as the cause of a vibration that he observed in his laboratory, but he was not the first to anticipate that earthquakes were associated with physical vibrations traveling through the solid Earth. In 1761, J. Michell was perhaps the first to suggest that earthquakes generated vibrational waves traveling in the Earth, and in 1807, Thomas Young and in the 1840s, W. Hopkins and R. Mallet also made similar suggestions. Around 1830, the French mathematician Simeon Denis Poisson had worked out the details of the theory of vibrations of an elastic solid. Poisson found that two distinct types of vibrations can be transmitted through a solid. The first kind is a compressional wave, in which the particle motion is parallel to the direction that the wave energy is traveling. The second is a shear or transverse wave in which the particle motion is perpendicular to the direction that the wave is traveling. Compressional waves can be transmitted through both solids and fluids, but shear waves can only be transmitted through a solid body, because a fluid has no restoring force in the transverse direction. These two wave types are sometimes called P and S waves, for Primary and Secondary, because the compressional (P) waves tend to travel faster and thus arrive at a seismic observing station before the S waves.

Following Von Rebeur-Paschwitz's epochal discovery, there was a flurry of activity to measure and record these seismic vibrations. He urged the creation of a worldwide network of seismographs. John Milne, working with the British Association for the Advancement of Science, implemented the first such network. Improved instruments were developed by Emil Weichert in Germany and Boris Galitzin in Russia. But geophysicists then faced the daunting task of interpreting the mass of data that resulted from these seismic networks, using it to infer the detailed properties of the interior of the Earth. As Agnew (2002, p. 6) put it:

> As these and other instruments were installed at observatories around the world, seismologists faced a new problem: sorting out the different "phases" involved, and relating them to different kinds of waves propagating inside the Earth. Theorizing about the Earth's interior was an active subject in the nineteenth century, but (rather like cosmology today) one in which a maximum of ingenuity was applied to minimal amounts of data. It was agreed that the depths of the Earth were hot and dense, but what parts were solid or liquid (or even gaseous) was the subject of much debate, though by the 1890's there was general agreement (from Kelvin and G. H. Darwin's tidal studies) that a large part of the Earth must be solid, and thus capable of transmitting the two types of elastic waves, along with the elastic surface wave proposed by Rayleigh in 1885. But it was not known which of these wave types would actually occur, and how much they might become indistinguishably confused during propagation. Given that the Earth was an inhomogeneous body, and that rocks were anisotropic, seismologists had a wide range of options available to explain the observations.

Weichert and his colleagues at Goettingen were involved in more than just instrumental development. The body of theory of waves in solids and fluids developed in the nineteenth century was available to be applied to the problem of the interpretation of the data from these networks of instruments. Herglotz and Weichert used mathematical techniques devised by Bateman to develop a solution (subject to some restrictive assumptions) to what is called the "geophysical inverse problem" in seismology. The inverse problem is easy to understand conceptually. If the Earth were a sphere whose internal physical properties were perfectly uniform throughout so that a seismic (vibrational) wave traveled at the same velocity everywhere, then calculating the time that seismic waves from a given earthquake would arrive anywhere on the surface of the globe would be a simple problem of simply calculating the chord (straight-line) distance through the Earth from the location of the earthquake and dividing by the velocity. But the wave velocities are not uniform through the Earth. They change with composition, state (solid or liquid), temperature, pressure, and density, all of which vary with depth in the Earth. Therefore, what is called the "forward problem" involves calculating the travel times of the waves from the earthquake to the point where they are observed with seismometers, based on a model of seismic wave velocities as a function of depth. And this forward problem is a fairly straightforward computation based on an integral equation. But, of course, we do not have a priori knowledge of the velocities as a function of depth, and so we cannot use the straightforward analytical techniques of the forward problem. Instead, we are able to measure the travel times themselves. The inverse problem is therefore the problem using these observations of travel times to deduce the changes in seismic velocity as a function of depth, in effect to "see" the elastic properties of the materials that comprise the inside of the Earth. And just to make things more complicated, we generally do not know exactly when and where the earthquake occurred, so these parameters also have to be deduced from the measured arrival times of the seismic waves, so we can then derive these travel times as a function of distance from the earthquake.

In one of the greatest triumphs in the earth sciences in the precomputer age, researchers collected the necessary data and then tackled and generally solved this inverse problem in the early decades of the twentieth century. First, as noted above, there was an immense effort that was started by Milne of constructing and deploying the instrumentation needed to measure the seismic vibrations in networks of observing stations distributed around the globe. Of course, the instrumentation had to be almost constantly upgraded as technological improvements to the instruments were developed. Not the least of the instrumental problems involved accurate timing—the clocks at the seismic observatories had to be carefully synchronized, using then-novel radio and telegraph capabilities. Then the colossal task of collecting and tabulating data from these instruments was also implemented with hand labor. And finally, the interpretation of those data in the solution of the inverse problem was a daunting task that required decades of hand computational effort.

There were many stunning discoveries that lay in reward for this enormous effort. Anton Mohorovicic discovered that there is a major discontinuity (a sudden increase) in seismic wave velocities at depths of about 50 kilometers beneath continents. It was later discovered that the same discontinuity occurs at depths of only a few kilometers under the ocean. The region above this discontinuity is termed the Earth's crust, and the region beneath is termed the mantle. And in 1908, Richard Oldham discovered that direct P and S waves disappear in a "shadow zone" at a distance of about 11,700 kilometers away from an earthquake. He further discovered that at larger distances the direct P waves reappear but the S waves do not. This odd pattern could be explained if Earth had a region with decreased P-wave velocities and no shear wave transmission below a depth that Beno Gutenberg eventually determined to be about 2,900 kilometers. The lower velocity of the P waves in this region will cause the waves to refract downward, away from the surface so that they are not seen in the shadow zone beginning at a distance of 11,700 kilometers, but they reappear still farther away (in the direction that they have been refracted). The lack of any reappearance of any S waves (that would have traveled through this region) indicates that the material in this region is a fluid, which cannot propagate S waves at all. This region below about 2,900 kilometer depth is called the core, and it is believed to consist largely of a molten nickel-iron alloy. And in 1936, the Danish seismologist Inge Lehmann succeeded in showing that there is a discontinuity in seismic wave velocities deep within the fluid core. This discontinuity within a fluid is now confirmed to be a fluid-solid boundary, and thus the very center portion of the Earth, now termed the "inner core," is believed to be solid iron, crystallized out of the fluid of the outer core. It is generally believed today that the heat released in the process of crystallizing this solid inner core is at least partly responsible for convective fluid flow within the core that generates the Earth's magnetic field (see the discussion in Song [2002, p. 927]). Computer technology turns out to be crucial to our ability to understand the dynamic processes in the fluid core that generate Earth's magnetic field.). We will have more to say about this magnetic field later.

These early (precomputer) determinations of the major structures of the interior of the Earth, its crust, mantle, core, and inner core represented a tour-de-force of observation, measurement, computation, and insight that involved massive amounts of hand labor. This pioneering effort was in many ways analogous to the voyages made by Columbus and other seafarers in the fifteenth and sixteenth centuries. Columbus's discoveries were enormously important for Europeans' knowledge of geography, but they represented only the beginning of a massive exploration of the newfound continents, of the discovery and mapping of their mountains, rivers, lakes, and other prominent features. Similarly, the identification of the major features of the interior of the Earth was only the beginning of a massive effort to explore the detailed structure of those features. And computer technologies were essential to this further exploration. The precomputer techniques were adequate to explore features for a simplified, radially symmetric Earth model, but exploring the details of more complex Earth models would require vastly greater amounts of data, quantities that could only be handled with computer techniques. According to Ray Buland of the U.S. Geological Survey (USGS), the USGS collects global seismic data at a rate of about three gigabytes per day (3×10^9 or three billion bytes per day), a quantity that no conceivable level of effort could handle in the absence of computer technology. And another seismic network, run by the Incorporated Research Institutions for Seismology (IRIS) consortium, collects even more data, about 16 gigabytes per day, according to Tim Ahern, the project manager for the IRIS data management system. Computer technology has caused a real sea change in seismology, according to Buland.

One of the most important changes in seismology in the computer era is that seismologists are now able to relax the assumption that the Earth is radially symmetric. The assumption of radial symmetry was good enough for establishing a reference Earth model, for determining structures at scales of thousands of kilometers such as the core-mantle boundary and the inner core boundary, and it was an essential feature of the calculations in the precomputer era because it provided an enormous reduction in the amount of calculation that was required. The assumption of radial symmetry meant that the (human) computer had to consider only the distance between the earthquake and the observing station and not the details of the direction as well. But with the calculating power of electronic computers it has now become possible to examine the variations in seismic travel times in azimuth (horizontal direction) as well as distance and to resolve real features in the Earth's mantle that were merely sources of error under the assumption of radial symmetry. Thus, seismologists today are able to resolve and observe features such as "mountain ranges" or large-scale undulations in the core-mantle boundary and temperature variations in the mantle that are associated with convective flow patterns. The technique of inverting massive amounts of data to determine seismic velocities across a grid of blocks that vary in azimuth as well as depth and distance is referred to as seismic "tomography," by analogy with CAT-scan imaging technology in medicine ("CAT" stands for Computerized Axial Tomography).

This seismic tomography is able to resolve features within the Earth's mantle at scales ranging down to a few hundred kilometers. This scale gives us a method that can be used to estimate the magnitude of the phase change in seismology associated with the introduction of computer technology. In the precomputer era, seismologists were able to detect radially symmetric features in the deep interior such as the depth to the core-mantle boundary whose scale is thousands of kilometers. Smaller-scale features, such as the structure of the crust and even smaller features such as oil and mineral deposits, could only be resolved at distances close to the surface (where oil and mineral deposits would be most useful, of course). Thus, the phase change that resulted from the introduction of computerized techniques is a change from detecting deep structures at scales of thousands of kilometers to structures at scales of tens to hundreds of kilometers, an increase of one or two orders of magnitude.

In addition, computers have allowed the development of entirely new approaches to seismology, including the analysis of the free oscillations of Earth. Again quoting Agnew (2002, p. 16):

> Much of what seismologists did with the WWSSN [the World-Wide Standard Seismograph Network, implemented in the 1960s] data would not have been possible without the other tool which became common at the time: rapid computation. As computing costs fell, and available funds rose, seismologists were able to speed up calculations which previously had taken up much time (epicenter location) and begin to do things which the labor of computation had ruled out before, such as compute surface wave dispersion in realistic structures. But the effect of the computer was not just to allow seismologists to model complex structures; it also gave them new ways to look at data. The ideas of signal processing and Fourier analysis, developed largely by statisticians and electrical engineers, began to make their way into seismology, to show what could be done with waveforms beyond timing them.
>
> The year 1960 brought an impressive demonstration of this new style of seismology, with the first detection of the Earth's free oscillations.

Again, conceptually, these free oscillations are not difficult to understand. When the Earth is shaken by an earthquake, it oscillates or "rings" like a bell. And the frequencies and amplitudes of the harmonics of these oscillations (or "normal modes" of vibration) can be analyzed to give us new information about the internal structure of the Earth. Computers were essential for detecting these normal modes and for implementing Fourier transform techniques to determine the frequencies and amplitudes of the harmonics of the observed surface motions. They were also essential to calculating the theoretical details of the oscillation modes that are expected from the known or hypothesized internal structure of the Earth. Comparison of these theoretical calculations with the observed spectra of the oscillations of the Earth led to massive computerized calculations to determine exactly what changes in the modeled structure of Earth were needed to provide a better fit to the observed harmonic oscillation data. This gave us additional information about the details of the Earth's internal structure. For example, conventional seismology allowed seismologists to determine that the outer core was a fluid but could not tell much about the properties of that fluid, particularly its viscosity (or "thickness," i.e., its resistance to flow). Detailed analysis of the Earth's observed normal modes allowed them to determine that the value of the viscosity was very small, roughly comparable to ordinary water rather than molasses, for example. And much of what we know about the physical properties of the solid inner core came from analysis of free oscillation data. Such data also allowed researchers to investigate what is called the "Q," the fractional energy dissipation in these harmonic vibrations within the Earth.

There are a number of other major applications of computer power in seismology. Just the calculation of hypocenter solutions (the location, depth, and time of the earthquake) for thousands of observed teleseismic events has contributed enormously to our knowledge of the detailed seismicity of Earth, the global distribution of earthquakes, as well as the detailed fault structures in earthquake-prone regions such as southern California.

The calculation of synthetic seismograms by reflectivity methods and mode-summing techniques has revolutionized our ability to interpret seismic data. If Earth did not have major discontinuities such as the core-mantle boundary, then calculating seismograms would be an easy, almost a trivial problem. But seismic waves are reflected from all the discontinuities in the Earth's internal structure, and all the reflections contribute to the observed vibrations at the surface. Calculating the effects of these reflections at varying incidence angles requires massive amounts of computation that are possible only with computer technology.

Also, the computation of surface wave dispersion curves for model structures requires an enormous computational effort. A dispersion curve expresses the change in wave velocity as a function of frequency, and the dispersion of an elastic wave is a strong function of the details of the structure through which the wave is propagating. It is not easy to determine the effects of small changes in the model parameters without massive computations. Extensive numerical calculations of the behavior of the surface waves in various models give important clues to the interpretation of the observed surface waves. Surface wave dispersion calculations were first done with massive amounts of hand calculations by Sezawa in the 1930s, before computers were available. The massive calculating power available with computers allows us to do these calculations today for Earth models that are much more complex and realistic than the ones that can be handled with hand calculations.

In addition, computers have enabled seismologists to calculate the details of the mechanisms and energetics of thousands of earthquakes. These calculations involve inverting the observed seismic waves to determine the orientation of the fault planes and the nature of the motions along those fault planes that produced the earthquakes in the first place, for example, was the motion horizontal (strike-slip faulting) or inclined at a sharp angle to the surface (normal and thrust faulting). Detailed knowledge of the directions of the fault motions helps to "see" and understand the stresses and strains that generate earthquakes and make estimates of the energy released. As Madariaga and Olsen (2002, p. 18) put it:

> Thanks to improvements in speed and memory capacity of parallel computers, it is no longer a problem to model the propagation of seismic ruptures along a fault, or a fault system, embedded in an elastic 3D medium. The enhanced computational power can be used to improve classical models in order to determine the grid size necessary to do reproducible and stable earthquake simulations.

The ability to use seismic waves and free oscillations to "see" the internal structure of Earth and the details of earthquake mechanisms clearly provided one of the major revolutions in our understanding of the structure and behavior of Earth, and computers have provided a massive increase in that capability in recent years. But another great paradigm shift in the earth sciences required a different set of phase changes. There is broad general agreement that the theory of plate tectonics, which explains many of the features of the surface geology of Earth, was one of the most important paradigm shifts in the history of earth science. This paradigm shift forced major revisions in ideas not only in geophysics but also throughout the rest of geology and even biology and evolution theory. But the original idea of continental drift met with strong resistance in the earth science community until two separate phase changes allowed us to see the vital evidence that supported it.

The earliest ideas about continental drift date back to production of the first accurate maps of the coastlines of the American continents, Africa, and Europe. A casual glance at such maps reveals a curious coincidence in the shapes of the continents across the Atlantic Ocean. They fit together like pieces from a colossal jigsaw puzzle. This unusual match in the shapes of the continents led to speculation that they had once been joined together. And in a series of now-classic papers written between 1912 and 1930, Alfred Wegener showed there was more to the story than a simple matching of coastlines. Wegener showed that not only did the shape of the continents fit, but the geology, both the fossil types and the crustal structures up to a point in time about a hundred million years ago, matched up as well. Wegener's ideas had a number of prominent supporters, including such men as Alex du Toit, who published in 1927 a detailed study of the remarkably close resemblance of Paleozoic and Mesozoic geology in Africa and South America.

But the idea of continental drift did not win universal or even wide acceptance for two basic reasons. Perhaps the most fundamental problem was the question of the nature of the physical mechanism that could move a continent some thousands of kilometers across the surface of the globe. Harold Jeffreys famously demonstrated that the force needed to push an object the size of South America through a presumed highly viscous substratum of rock was vastly larger than any known physical mechanism could provide. The major problem here was that Wegener's model assumed that the substratum of rock was not moving. It was not realized until the middle of the twentieth century that the substratum was also moving, that the Earth's mantle is subject to convective motions. The metaphor that is conventionally employed by geophysicists is that a continent, instead of plowing through the mantle like a raft on a lake, is actually riding on the motions of the mantle below like a box on a conveyor belt.

But the other reason for resistance to the idea of continental drift is that geologists at the time had a strong and perfectly reasonable prejudice in favor of vertical motions of continents over horizontal motions. This prejudice was reasonable because they could see firm evidence for vertical motions. The presence of oceans on this planet, and, in particular, the existence of sea level, provided a natural asymmetry that gave geologists a means to observe vertical motions in the past but not horizontal motions. Geologists had known for centuries about fossils that were clearly of marine origin but were found at locations near the tops of mountains, several kilometers above sea level and far beyond the range of any conceivable change in sea level. Clearly, these rocks had been moved vertically by thousands of meters from their origins below sea level to their present locations on mountaintops. So geologists had long known about and had become comfortable with the idea that large blocks of rock could move thousands of meters in the vertical direction. In contrast, there was no evidence that was yet recognized or accepted for horizontal motions, apart from the coincidence of land shapes and geology across the Atlantic. And this asymmetry in the nature of the observed data led to a preconception that tectonic motions occurred primarily in the vertical direction rather than the horizontal. Geologists had made a classic blunder of mistaking the absence of evidence (for horizontal motions) for evidence of the absence of such motions.

Therefore, to explain the similarities in the geology across the two sides of the Atlantic that Wegener and others had observed, geologists assumed that they were due to vertical motions of the Earth's crust, not the horizontal motions proposed by Wegener. It was hypothesized that the continents on both sides of the Atlantic had once been connected by what were termed "land bridges," ocean-spanning land masses that had recently sunk to the bottom of the ocean like the legendary Atlantis. This explanation required only a few kilometers of vertical motions by the land bridges, an idea that geologists found much more congenial than the horizontal motions of thousands of kilometers that were required by Wegener's hypothesis. Thus, the idea of continental drift was well ahead of its time, ahead of the phase changes that would allow geologists to see convincingly the horizontal motions of the continents in the past.

The phase change that gave scientists the ability to see the horizontal motions that continents had undergone in the past came from an unlikely source, the study of rock magnetism. Magnetism was diagnostic because the Earth has a magnetic field, and some rocks carry small amounts of magnetic minerals such as magnetite (Fe_3O_4) that can respond to that field and, in effect, fossilize a weak magnetization of the rock that records the direction of the Earth's field. One of the most important mechanisms of inducing a magnetization in rocks is called "thermal remnant magnetism." Thermal remnant magnetism occurs when melted rock cools through a temperature known as the Curie point, the temperature below which permanent magnetization becomes stable. For magnetite the Curie point is a temperature of about 670 degrees Celsius. At this temperature the magnetite in the rock will be "imprinted" with a magnetization that corresponds to the Earth's field at that time. In the early part of the twentieth century, Bernard Brunhes was able to show that contemporary lavas in Iceland are magnetized in the direction of the Earth's field, and earlier lavas are magnetized in the same direction or the reversed direction, and that still older lavas are magnetized in directions that are at large angles to the current field. This work contained the germ of several major discoveries, including the fact that magnetism in rocks could be used to infer the Earth's field in the past, that the Earth's field flips direction at irregular intervals, and that the continents have drifted relative to the magnetic pole in the past. But Brunhes's work did not receive the attention it merited. Hallam (1992, p. 163) notes that "it is remarkable that such fascinating and obviously significant phenomena should have been so widely ignored for so many years."

In the 1940s, P. M. S. Blackett at Cambridge developed a very sensitive magnetometer, an instrument that can measure magnetic fields. Blackett's students, especially S. K. Runcorn, K. M. Creer, and E. Irving, were able to exploit this new instrument to investigate the Earth's magnetic field in the past by measuring the remnant magnetization of rocks with unprecedented precision. This magnetization could not only determine the direction to the magnetic pole like a fossilized compass needle, as Brunhes had noted, but it could also determine the distance to the pole by measuring the dip angle of the field, the angle to the horizontal. At the magnetic pole the dip is 90 degrees—the field there is straight up and down; conversely, at the magnetic equator, halfway between the north and south magnetic poles, the dip is zero—the field is perfectly horizontal. And between these extremes the dip takes on intermediate values that can be used to estimate the magnetic latitude, the angular distance to the magnetic pole. Runcorn found that the derived magnetic pole positions drifted with time, suggesting that either the pole was drifting or the continents were or both. And by about 1950, he was able to show that the track of the pole position over time as measured from rocks in Europe was different from that shown by rocks in North America. Furthermore, he showed that the difference vanished if you assumed that the American continents had once been close to Europe and Africa and had drifted away. This was one of the first completely novel pieces of evidence in favor of Wegener's hypothesis, but it was not widely accepted at the time. As Hallam noted (1992, pp. 174–175): "There was still a good deal of suspicion of the new and unfamiliar research technique, and the reliability of its results were widely questioned. In North America the new palaeomagnetic work was generally received with extreme skepticism."

Magnetic observations would later provide a second piece of critical evidence for the plate tectonics paradigm shift in the 1960s. But before we describe the newer magnetic observations, we need to examine a third phase change in the earth sciences, a phase change associated with the ability to see the ocean floor.

We often forget how little was known about the floor of the deep oceans until very recently. Even something as simple as the shape of the ocean floor was a complete mystery until the middle of the twentieth century. Prior to the development of modern electronic instrumentation little was known about the depths of the ocean basins except in their most shallow parts close to shore. The problem is that there was only one known way to determine the depth of the ocean: Tie a weight to the end of a rope and then drop it to the bottom. Typically, the length of the rope would be measured as it was drawn back up. This technique was referred to as "sounding" the depth of the water, and it had been in use from antiquity. But sounding was largely limited to shallow water because ships seldom carried the miles of line and other special equipment that would be required for deep-ocean sounding.

In the later part of the nineteenth century, the Royal Navy's HMS *Challenger* was sent out on a famous expedition to explore the oceans and make heroic efforts to measure the depth of the central ocean basins using conventional sounding techniques. The crew of the *Challenger* experimented with using 12 miles of piano wire for depth sounding. But the piano wire did not work very well because it tended to kink as it was paid out. They then resorted to using miles of conventional hemp rope wound on a special drum winch that had been designed to handle such quantities of line. In its four-year voyage, the *Challenger* managed to sound the ocean at 362 observing stations, including one location that is still called the "Challenger Deep," a 26,850-foot depth in the Marianas Trench in the western Pacific.

In 1895, John Murray accurately described the *Challenger* expedition as having produced "the greatest advance in the knowledge of our planet since the celebrated discoveries of the fifteenth and sixteenth centuries." Yet, heroic as it was, *Challenger*'s 362 depth values distributed around the global ocean fell far short of producing anything like a complete picture of the shape of the ocean floor. And the cost of the expedition was such that, given an absence of compelling military necessity for the Royal Navy, it was unlikely ever to be repeated. The only other information about depths at midocean locations that was available in the nineteenth century came from the laying of transoceanic telegraph cables, beginning in the 1850s, a novel and unusual method of sounding. The ships laying the cables discovered that there were mountains in the middle of the Atlantic Ocean, but again due to the restricted areal coverage available from the cable-laying efforts, the full extent of the mid-Atlantic ridge was not discovered.

The real phase change in our ability to observe the depths of the central ocean occurred with the development of electronic depth sounders in the 1940s. These depth sounders measure the travel time of sound waves that are reflected from the bottom. The technique was called sonar (SOund Navigation And Ranging), by analogy with radar (RAdio Detection And Ranging) for the similar measurement technique that used radio waves instead of sound. Radar could not be used to observe the ocean bottom because radio waves do not penetrate seawater. But with sonar equipment a ship could measure the depth of the water underneath it effortlessly and continuously as it transited the various oceans of the world. For regions of the ocean that have a lot of ship traffic, such as the North Atlantic, these ship tracks were sufficiently dense that reasonably complete maps of the ocean bottom were quickly assembled. And stunning discoveries were quickly made. By the 1950s, Bruce Heezen at Columbia University was able to show that a mountain range ran all the way down the center of the Atlantic from Iceland clear around the tip of Africa and extended into the Indian Ocean.

Opponents of Wegener's continental drift ideas had argued that the matching of the coastlines on both sides of the Atlantic was merely an interesting coincidence. But the presence of a mountain range in the middle of the ocean that also matched the shape of the shorelines on either side of it was a serious blow to the idea that the coastline match was simply a coincidence. And Heezen interpreted a canyon that was found to run down the center of the mid-Atlantic ridge as a tensional feature. Apparently, something was tearing the entire mountain range apart, right down the middle.

A second important discovery from the depth-sounding measurements was that the deepest points on the ocean floor, such as the Challenger Deep, were found in narrow trenches that ringed the edge of the Pacific Ocean. In the early 1960s Harry Hess at Princeton tied many of these newly observed facts together in what came to be called the "sea-floor spreading" hypothesis. He built on some ideas that had been originally proposed by Arthur Holmes in an address in Glasgow in 1928. Hess proposed that the sea floor was controlled by convective flow in the Earth's mantle. In Hess's model, the sea floor was created by cooling magma from the mantle that was upwelling by convection at the midocean ridges. The newly created ocean floor then spread outward from the ridges toward the ocean trenches such as the Marianas Trench that had first been sounded by the *Challenger* expedition. Hess and Holmes both suggested that this process provided the motive force for continental drift—that the continents were carried along like passive blocks on the motions of the underlying mantle. As Hess explained it (quoted in Hallam, 1992, p. 167):

> The mid-ocean ridges could represent the traces of the rising limbs of convection cells while the circum-Pacific belt of deformation and volcanism represents the descending limbs. The Mid-Atlantic ridge is median because the continental areas on each side of it have moved away from it at the same rate. . . . This is not exactly the same as continental drift. The continents do not plough through oceanic crust impelled by unknown forces, rather they ride passively on mantle material as it comes to the surface at the crest of the ridge and then moves laterally away from it.

Hess's hypothesis contained some of the critical elements that led to the modern theory of plate tectonics. It tied together many disparate facts about the continents and oceans and at the same time it removed the principal objection to Wegener's ideas, the difficulty of pushing a continent through a passive underlying mantle. And it soon received solid confirmation from a series of magnetic measurements in a way that ties our two story lines back together. This confirmation connects the phase change caused by magnetic measurement instrumentation with the one caused by sonar measurements of the ocean depths.

In the early 1960s, Fred Vine at Cambridge and Lawrence Morley of the Canadian Geological Survey were both studying some peculiar patterns of magnetic anomalies that had been observed in the floor of the Indian Ocean. These measured anomalies are simply the observed magnetic field after removal of the large-scale regional field that is dominated by the overall field of Earth. The magnetic anomalies on the sea floor were observed to occur in stripes that ran parallel to the Carlsberg ridge. Early attempts to interpret this peculiar magnetic striping hypothesized compositional differences on the ocean floor (the chemical composition of the rocks that form the ocean floor was sampled only erratically and was very poorly known) or thermal or even topographic differences.

Vine and Morley independently realized that Hess's ideas about sea-floor spreading could be combined with the fact that the Earth's magnetic field reverses itself at irregular intervals of about a million years (as first noticed by Brunhes a half-century earlier), to provide a simple and elegant explanation of the observed striping of the rock near the midocean ridge. As the sea-floor was created from melted mantle rock at the ridgeline (according to Hess's theory), it would acquire a remnant magnetization in the direction of the Earth's field at the time. The rock would then move by sea-floor spreading away from the ridge in a magnetized stripe that was oriented parallel to the ridge, preserving that particular direction of remnant magnetization in the rocks of the sea floor. Then when the Earth's field switched directions, the new sea-floor at the center of the ridge would acquire the new magnetization direction and begin to produce a new stripe that was magnetized opposite to the earlier stripe. The magnetized stripes that moved away from the ridgeline have been described as being analogous to the tape in a gigantic magnetic tape recorder. The moving sea-floor preserves the magnetization that was imprinted on it when it solidified, much as the tape in a conventional tape recorder preserves the magnetization that was imprinted on it when it passed over the record head of the tape recorder.

Vine and Matthews's 1963 paper on the subject was published in *Nature* and quickly became famous. In an ironic twist, Morley's paper with largely identical results was also submitted to *Nature* in 1963 but received an unfavorable review and was not published. Morley's article was eventually published in another journal, but, in Hallam's words (1992, p. 170): "Thereafter his involvement with the subject ceased." Vine, on the other hand, went on to make critical measurements, showing that the observed magnetic striping had the symmetry that would be expected under the sea-floor spreading hypothesis. In other words, the stripes matched on opposite sides of the ridge because the spreading is symmetric in both directions away from the ridgeline.

The theory of plate tectonics is clearly one of the great paradigm shifts of modern earth sciences and it had yet another critically important confirmation that depended on the phase changes that preceded it. Two of the earlier phase changes are also involved in this next piece of the plate-tectonics paradigm shift: the revolution in seismology that allowed us to use the elastic vibrations caused by earthquakes to see the structure and behavior of the Earth, and the revolution in depth sounding that allowed us to see the shape of the ocean floor. This narrative begins with a simple observation about the shape of the ocean floor. Although on a global scale the midocean ridges form a nearly continuous mountain chain that runs for tens of thousands of kilometers along the ocean floor, on a much finer scale there are small breaks or discontinuities in the range. In particular, there are many places where the midocean ridge is displaced by tens of kilometers along what is clearly a fault line, a break in the rock floor of the ocean. These fault lines show up as linear features on the ocean floor that are called "fracture zones." It was J. Tuzo Wilson in Toronto who had the next critical insight about the importance of these fault lines. Wilson saw that these faults would behave very differently if the midocean ridges were centers of sea-floor spreading than if they were more conventional (and passive) mountain ranges as Heezen and others suggested.

If the midocean ridges were passive mountain ranges, then the fault zone should show earthquakes all along the fault because the rock on either side of the fault is moving in opposite directions, as shown in Figure 1 ("Transcurrent Fault"). However, if the midocean ridges are spreading centers, then the rock motion on either side of the fault is completely different. In particular, on the sections of the faults that are outside the two parts of the midocean ridges (far left and right in the figure), the rock is moving in the same direction on both sides of the fault, and no earthquakes should be expected there. Only in the portion between the two ridges is the rock moving in opposite directions. Therefore, this is the only portion of the fault that should produce earthquakes. Wilson also noted that the sense of the rock motion in this region between the ridges would be in the opposite direction from what it is in the other model of Figure 1. Lynn Sykes at Columbia University was able to use modern seismic measurements to find very pre-

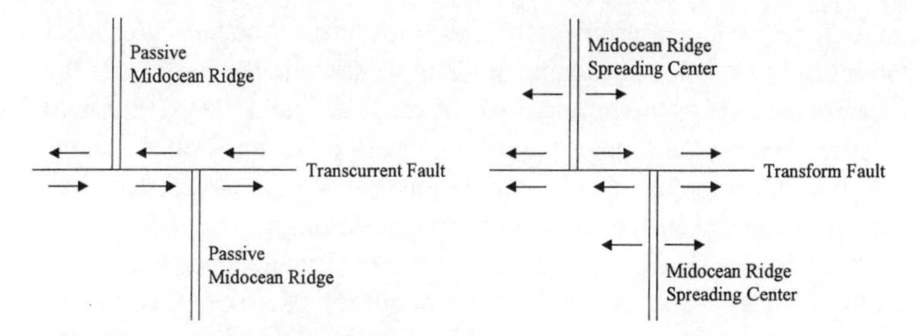

Figure 1. Transcurrent and transform faults

cise locations of earthquakes near the midocean ridges and show that the earthquakes are found only on the portion of the fracture zone between the ridgelines, exactly as predicted by the sea-floor spreading hypothesis. Further detailed analysis of the seismic data showed that the rock motion is in the correct sense for the sea-floor spreading model and not correct for the other model. Wilson called these faults that separate spreading ridges "transform faults." The other type of fault in the figure is sometimes called a "transcurrent fault." Wilson's and Sykes's work was regarded by many as the most convincing evidence yet for sea-floor spreading. It was direct and incontrovertible evidence showing the direction of motion of the sea-floor and showing that the midocean ridges are, in fact, the centers of sea-floor spreading.

Thus, the paradigm shift associated with the development of plate tectonics and the related development of a detailed picture of the internal structure of the Earth was ultimately based on three phase changes associated with the development of seismology, magnetometry, and sonar measurements of the depths of the ocean. And we have seen how computer technology has produced a new phase change in seismology leading to seismic tomography as well as to measurements of the Earth's free oscillations, producing vast new improvements in our ability to "see" the interior of the Earth. But how is computer technology affecting our ability to measure the ocean depths and to measure the Earth's magnetic field? There are several answers to these questions, but perhaps the most obvious involves measurements made from instruments on satellites in orbit around the Earth.

One of the simplest measurements that can be made from a satellite is simply its altitude. Laser and microwave radars can measure the distance from the satellite to the Earth's surface below and particularly the distance to the surface of the ocean. It might not be obvious at first, but such measurements of the ocean surface can tell a great deal about the ocean floor beneath for a simple reason: The surface of the ocean responds to the gravity field of the Earth, and the gravity field is perturbed by the topography of the ocean floor. In essence, undersea mountain ranges have a gravitational attraction that distorts the shape of the sea surface above the mountain range. The surface displacements are small, usually only a few centimeters when the topography of the ocean floor changes by hundreds of meters. But the altimeters aboard the satellites are able to detect the changes in the sea surface at this level.

And because the satellites are able to make measurements across the entire ocean surface, the coverage of the new maps of the ocean floor from satellite altimetry is essentially complete, much more complete than was ever possible from sonar measurements taken from ships, especially in regions of the oceans that are remote from shipping lanes and the conventional sonar measurements are therefore rare. Thus, although the ocean floor was almost totally unknown at the middle of the twentieth century, today there are no major unknown geographic features left on the sea floor. One of the maps of the ocean floor derived from such satellite altimetric measurements can be found on the Web at http://www.ngdc.noaa.gov/mgg/fliers/97mgg03.html. In this map the network of midocean ridges is clear, as is the line of trenches around the Pacific Rim (and one single trench in the Atlantic Ocean north of Puerto Rico). The great fracture zones that mark the transform faults are also clearly seen, particularly in the Pacific region.

This map could not have been constructed in the precomputer era. Not only are computers essential for the design, launch, and operation of the satellites involved, but also to process the altimeter data requires extremely precise knowledge of the location of the orbiting satellites. Knowledge of the satellite location requires computerized processing of enormous volumes of tracking data to adjust the numerical integration of the orbit. Then when the satellite position is known, there is a plethora of corrections that need to be made to the raw altimeter range data, corrections for refraction delays in the atmosphere, tidal motions of the ocean surface, wind, and current effects and other perturbations.

Modern measurements of the Earth's magnetic field are also made from satellites, and again an enormous amount of computer calculation is needed to process the resulting data. Not only is the orbit determination problem highly computer intensive, but also magnetic observations have to be corrected for effects such as electrical currents in the Earth's ionosphere. Also, in the case of magnetic observations from space, there is another version of an inverse problem. The forward problem in magnetics would involve having a priori knowledge of the magnetic properties of the Earth itself and then using that information to calculate the magnetic field at the location of the satellite. Of course, in the real world that is not what happens—you make measurements of the magnetic fields at the location of the satellite and then the inverse problem involves using these field observations to calculate the magnetic properties of Earth below. This inverse problem is computationally intensive and subject to a variety of difficulties, including nonuniqueness, numerical instabilities, and other problems that seriously complicate the calculations. Nevertheless, important data about the magnetic properties of the Earth's surface can be obtained from massive calculations based on the satellite magnetic observations. And these observations are not confined to Earth. Orbiting magnetometers have given important evidence about the internal structures of the Moon and other planets.

Altimetry and magnetic observations can also be done from aircraft, and for regional studies, aircraft measurements are frequency cost-effective compared to the costs of satellite operations. But processing the data from aircraft observations has all the problems of processing the data from satellites plus the added problem that, until recently, it was impossible to know the aircraft location as accurately as the satellite positions were known. But with modern computer technology it is possible to use the Global Positioning System of satellites to determine aircraft locations with an accuracy in the range of a few centimeters, sufficiently accurate for both altimetric and magnetic observations.

It is not easy to quantify the phase change associated with satellite and aircraft observations of the sea floor or of the Earth's magnetic field, but it is clear that the amount of observing that can be done, the amount of data that can be obtained about the Earth, is orders of magnitude greater than could be obtained without computer power. So again in the earth sciences we see the familiar pattern that we saw in the other sciences, that novel observing techniques including seismology, sonar depth measurements, and magnetic observations create a set of phase changes that allow us to see things that could not have been seen before. (A detailed discussion of these topics and many other novel and important applications of computer technology can be found in Lee et al., 2002.) These phase changes have produced a major paradigm shift in the form of plate tectonic theory. And computerized observing techniques are generating still greater phase changes today that are allowing us to explore the structure of the interior of the Earth as well as other planets in far greater detail than was ever possible in the past.

7

PHASE CHANGES IN METEOROLOGY

Meteorology, the study of the weather and the behavior of the Earth's atmosphere, has been transformed by computer technology perhaps more than any other field of science. Today computers are essential both for observing the atmosphere and for interpreting those observations in modeling and predicting the weather. And much like mathematics, there is no single invention in the past whose impact on meteorology is as pervasive as the computer, nothing comparable to the telescope in astronomy or the microscope in biology. (Orbiting satellite observations, which indeed revolutionized meteorology, are so intimately linked to computer technology that they will be considered part of the computer revolution in the discussion here.) Instead, there has been a succession of inventions of meteorological instruments such as the barometer and the thermometer that were important in their own right but which did not immediately revolutionize the field. This is in large part because meteorology is so vast and complex a subject that no single invention prior to the computer had much more than local or regional impact. There are too many complex facets of the field of weather science for all of them to be affected at the same time by any single invention in the past. Indeed, one of the most remarkable things about computer technology is that it has revolutionized all aspects of meteorology at the same time, just as it has in many other fields that we have explored. And in contrast to mathematics, the impact of computers is universally recognized by meteorologists. I know of no case in which meteorologists have expressed any reluctance to using computer technology.

Like the earth sciences, the study of the weather goes back to antiquity. A general interest in the weather is even more widespread than an interest in astronomy or earthquakes. Virtually everyone is affected by the weather, and in agrarian societies that were dependent on farming for subsistence, the weather was often a life-and-death matter on a daily basis. It is probably safe to say that more casualties have been caused by bad weather, by drought and flood, than by earthquakes and volcanoes combined.

The study of meteorology has been punctuated by inventions such as the thermometer and the barometer that allowed researchers to see or observe phenomena that could not be seen or could not be seen as well without them. And just as astronomers are able to make some observations using only the naked eye, without using telescopes and other instruments, weather researchers are able to observe many of the properties of the weather, including winds, clouds, and temperatures, using only their unaided senses. Indeed, this was the only way to observe many weather phenomena prior to the seventeenth century. But although the human body can sense such atmospheric properties as temperature, it does not generally sense them quantitatively. And it has only a very limited ability to sense important properties such as barometric pressure and relative humidity.

One of the most important meteorological instruments, the barometer, was invented in the early seventeenth century. The full history of the invention of the barometer is highly complex and parts of it are quite controversial. (Knowles Middleton [1969, pp. 3–42] gives a detailed discussion.) The barometer was invented for purposes that had nothing to do with meteorology, much as the first seismometer was invented for purposes that had nothing to do with geophysics; indeed, the meteorological implications of the barometer were not even suspected until after it was developed. The original impetus for experimenting with barometers centered on an investigation of the nature of the vacuum. Early barometers consisted of tubes of glass filled with a fluid, generally water or mercury. If such a tube were set up in a vertical position with the top end sealed off and the open (bottom) end in a basin of fluid, it was found that the fluid would not fill the entire tube. Instead, the upper level of fluid would fall to a fairly constant level that depended on the density of the fluid, about 30 inches for mercury and about 32 feet for water. The region above the fluid in the closed tube was a vacuum, in principle a perfect vacuum, except for the vapor pressure of the fluid and the presence of trace gas impurities. But the existence of a vacuum was a subject that church authorities, following Aristotle, regarded as anathema (Knowles Middleton, 1969, p. 15). Many of the problems with our present understanding of the history of the barometer resulted from the fact that researchers were afraid to publish their results. Thus, Galileo claimed to have invented the barometer, but he did so only in private correspondence that was not published until much later. He had had enough trouble with his publishing of Copernican ideas.

Evangelista Torricelli, who had worked with Galileo, was the first to explain the height of the fluid in the glass tube in terms of a balance of the pressure of the fluid against the pressure of the atmosphere outside the tube. He was also the first to report that the height of the fluid (and thus the pressure of the outside air) fluctuated from day to day. Again, Torricelli's work is described only in private correspondence that has survived (Knowles Middleton, 1969, pp. 11–15). The connection between the changes in atmospheric pressure and changes in the weather was not recognized until later. Descartes was apparently the first to attach a measuring scale to observe and record the height of the fluid in a barometer, as well as the first to suggest that the relation between changes in barometric pressure and changes in the weather should be investigated (Knowles Middleton, 1969, p. 23).

Another instrument, the thermometer, was also developed in the early seventeenth century. Temperature is one of the easiest meteorological parameters to sense without instruments. Indeed, maintaining a constant temperature is so important to any mammal's metabolism that we have evolved a specific sense to measure temperature. But our temperature sense is not quantitative—we can broadly tell hot from cold and comfortable from uncomfortable, but we cannot easily discern whether the temperature is 20 or 22 degrees Celsius.

Early thermometers did not face the difficulties of opposition from the church that impeded the development of telescopes and barometers. But they faced a more subtle problem in that the very nature of temperature was misunderstood for centuries. Temperature and heat were thought to be the same thing, and that "something" was thought to be a sort of fluid, sometimes called phlogiston, that flowed into and out of material bodies. The real nature of temperature and heat involved some of the deepest and most difficult problems in physics, many of which were not sorted out until the development of thermodynamics and statistical mechanics and even quantum mechanics in the nineteenth and twentieth centuries.

But reasonably accurate thermometers were developed long before the nature of temperature was well understood. The physical phenomenon that underlies the functioning of most thermometers to the present day is the fact that the volume of most physical bodies, including fluid bodies, changes linearly with temperature. Early thermometers depended on the change in the volume of air contained in a bulb that had fluid at the bottom. The familiar liquid-in-glass thermometer was apparently first proposed by a doctor named Jean Ray in about 1630. By the mid-1650s, thermometers with reasonably modern appearance were being constructed; many of these can still be seen in the Museo di Storia della Scienza in Florence (Knowles Middleton, 1969, pp. 51–52). In the later part of the seventeenth century, Huygens and Hooke began to address the problem of calibrating thermometers, using the freezing and boiling points of water, and by the early eighteenth century, Fahrenheit, Celsius, and Roemer had constructed temperature scales that are still in use today (Knowles Middleton, 1969, pp. 57–58).

Instruments to measure humidity are called hygrometers, and according to Knowles Middleton (1969, pp. 84–85), they can be classified into five different types:

> In order of their first invention, they are as follows: (1) those depending on the hygroscopic properties of various substances (hygroscopic hygrometers); (2) those depending on the formation of dew on a surface that can be artificially cooled (dew-point hygrometers); (3) those depending on the reduction of temperature produced by evaporation from a moist surface (psychrometers); (4) those in which the moisture is taken from a known volume of air and weighed or otherwise measured (absorption hygrometers) and (5) miscellaneous instruments depending on other physical properties.

As with the barometer and the thermometer, the invention of all these types of hygrometer is a rather complicated story that Knowles Middleton describes in some detail. Measurements today are predominantly made with type 2 and 3 instruments, but a set of instruments of type 5 has become important today. These instruments depend on the properties of emission or absorption of electromagnetic radiation by water vapor. This class of instruments is particularly important for meteorological observations from orbiting satellites that will be discussed later.

Other instruments for measuring properties such as wind direction and speed, rainfall, and cloud heights were developed through the seventeenth and eighteenth centuries, so that by the middle of the nineteenth century, meteorologists had a variety of instruments at their disposal for measuring the properties of the atmosphere. But their ability to "see" meteorological effects was hamstrung by an inability to transmit data from one location to another and, particularly, by an inability to transmit the information rapidly to a central facility for processing the data in time to make useful predictions. This is why the invention of modern electronic communication, beginning with the telegraph in 1846, was particularly important to meteorology. As Monmonier (1999, p. 39) noted, "Meteorologists quickly grasped the significance of the electric telegraph." Joseph Henry, the secretary of the Smithsonian Institution, organized a network of weather stations that reported daily by telegraph. By 1860, the network included 45 stations, extending from Vermont to Iowa and Louisiana. But the start of the Civil War cut the network ties to stations in the Confederacy, and a fire in the Smithsonian building in 1865 ended the project.

Meanwhile, in Europe researchers were also aware of the implications of telegraphic communication for meteorology. According to Monmonier (1999, p. 42), in 1848, "John Ball addressed the British Association for the Advancement of Science with a plan for telegraphic storm warnings. Noting that 'atmospheric disturbances' moved no more rapidly than 20 miles an hour, Ball suggested that 'with a circle of stations extending about 500 miles in each direction, we should in almost all cases be enabled to calculate the state of the weather for twenty-four hours in advance.'"

But the major deficiencies of these early efforts were clear to most researchers. There were not enough stations reporting, and the collection of data in a timely fashion was still difficult and expensive. Also, three-fourths of the globe is covered by oceans, and data collection and coverage from locations at sea were particularly difficult. And these efforts faced still another difficulty that was less obvious. All the measurements were taken at ground level, but, of course, many important weather phenomena, especially storms, occur at elevations as much as several miles above the surface. In mountainous regions, it was possible to take measurements at a variety of elevations above sea level, but mountains were not conveniently located everywhere that measurements were desired, and the mountains themselves affected the weather conditions that were being measured. In the late nineteenth century, there were only two general approaches to the problem of making weather observations above the Earth's surface: the use of balloons and kites. Both methods may seem simple and obvious today but in an era that preceded the invention of synthetic fibers and plastics, of nylon and kevlar fiber and polyethylene and mylar films, neither kites nor balloons were as simple to fabricate and operate as they are today.

Nevertheless, attempts were made to use balloons for weather-observing platforms, beginning in the late eighteenth century and continuing through the nineteenth (Knowles Middleton, 1969, pp. 287–290). Balloons had one very serious defect: unless the balloon was manned or tethered, it tended to drift away and get lost. And the loss of a balloon meant more than just the loss of expensive instruments. In the era before radio telemetry was available, it also meant the loss of all recorded data.

Alexander Wilson used kites to measure upper air temperatures in Scotland as early as 1749. Wilson used a set of burning fuses to release thermometers from a kite at specified times and a primitive sort of parachute made of paper tassels that let the thermometers fall to the ground intact. It was assumed the reading of the thermometer did not change significantly as it fell to Earth. Apparently, no record survives of Wilson's resulting temperature measurements (Knowles Middleton, 1969, pp. 291–292). Much of the following century was spent developing a variety of instruments that would record temperatures and barometric pressures while suspended on kites.

In 1893, Lawrence Hargrave published a description of a box kite. According to Knowles Middleton (1969, pp. 292–293): "The advantages of this construction, which can be made to fly more steadily, and at a higher angle, were immediately appreciated, and it may fairly be said to have revolutionized meteorological kite flying, to which a great deal of attention was being paid at the time especially in the United States." By 1895, kites were being flown at the Blue Hill Observatory in Massachusetts, carrying instruments to record temperature, barometric pressure, and relative humidity, and kites are still in use for specialized applications (Balsley et al., 1998).

But kites and balloons had serious problems as weather-observing platforms. The main problem, of course, is that there were never enough of them to observe any significant fraction of the atmosphere. Equally serious, in an era that preceded electronic telemetry, the data from these platforms could not be accessed until the kite or balloon returned to the ground. With the invention of electronic telemetry the usefulness of balloons expanded enormously, and the use of unmanned balloon "radiosondes" that transmit meteorological data in real-time to ground stations below remains an important method of collecting meteorological data today.

The invention of manned aircraft in the twentieth century vastly expanded our ability to sample large regions of the atmosphere, especially in regions of the globe where air traffic was common, and indeed, measurements from aircraft are a major component of meteorological observations that are available today. But even in the era of frequent and common transcontinental air traffic, the coverage of the globe from such measurements is spotty at best, and almost nonexistent in areas such as the southern oceans or south polar regions where flights are uncommon.

The development of radar during World War II provided another novel technique for measuring atmospheric parameters, especially cloud locations and movements. Early radars operated without computers, but modern weather radars are heavily dependent on computer technology and are a significant part of the phase change in meteorology that is being generated by the computer revolution. Modern Doppler weather radars, that measure not only the positions of clouds but also their velocities, require computers to process the frequency and time of the returning radar signals in order to provide the necessary observations of the velocity as well as the position of the targets (clouds) being observed with radar.

Thus, at the dawn of the computer era, meteorologists had developed a variety of techniques for "seeing" weather phenomena. And they had deployed these instruments across broad areas of the land and more erratically at sea and in the air. But the atmosphere is so vast a system, and its behavior is so complex and chaotic (in a technical sense of that term) that at the middle of the twentieth century, effective understanding and prediction of weather patterns was little improved over what had been done with simple observations of clouds and winds millennia ago. Weather prediction was sufficiently erratic to be a common subject for jokes such as "I just shoveled two feet of 'partly cloudy' off the driveway."

But the computer era ushered in two major changes in addition to radar that completely revolutionized our ability to see and understand weather patterns. First, the introduction of computerized observations from orbiting satellites provided fairly uniform global coverage of meteorological observations over land and sea for the first time. Second, computers gave us the ability to perform the massive calculations needed both to assimilate the vast quantity of data being provided by modern observing systems and to handle the calculations needed to model the behavior of the atmosphere.

Computers are an important component of modern weather radar technology, but they are absolutely vital to weather satellites. Weather-observing satellites simply could not exist without computers. The design, launch, and operation of weather satellites (not to mention data collection and analysis) are completely dependent on computers. One of the first and most important functions of these weather satellites was to take pictures of cloud formations. These satellites would thereby measure cloud coverage and cloud motions on a global basis (land and sea) for the first time, using computerized digital imaging techniques. But weather satellites can do much more than just take pictures of clouds. They can observe most of the atmospheric parameters such as temperature and pressure that meteorologists need for modeling the behavior of the atmosphere, parameters that had previously been available only from ground stations and from vehicles such as ships, kites, and aircraft.

The full theory of sensing atmospheric parameters from satellite observations is a subject that can only be sketched here. Basically, the chemical and physical state of the atmosphere produces a variety of electromagnetic effects, both radiation and absorption, that can be observed and measured from satellites overhead. For example, water vapor has a characteristic emission signature at a frequency of about 22 gigahertz (22 billion cycles per second). In effect, water vapor has a "color" of 22 gigahertz that we would be able to perceive with our own eyes if they were sensitive to that frequency band. Of course, eyes are sensitive only to a narrow band of frequencies in the optical band, much higher than 22 gigahertz, but it is not difficult to build receivers that can detect and measure radiation in this band. And temperature and barometric pressure also produce characteristic radiation signatures that can be sensed from satellites. Barometric pressure, for example, will broaden the width of the characteristic spectral lines of molecules. One critically important meteorological parameter, the velocity of the wind, is perhaps the most difficult parameter to sense remotely from satellites. Lidar technology (laser radar) is being developed to address this problem. According to Baker et al. (1995, p. 885): "The technology to deploy a space-based Doppler wind lidar is now available. The deployment of such an instrument would fundamentally advance the understanding and prediction of weather and climate."

The sensing of atmospheric parameters from satellite platforms constitutes a set of very difficult problems that are an area of very active research today, and an area where numerous technological improvements can be expected in the near future.

The interpretation of these remote sensing data from satellites is a very complicated problem, but modern computers can handle the extensive calculations, parameter fitting, Fourier transforms, and other analysis techniques that are needed to interpret the satellite data. Computerized data from satellites can therefore give us reasonably uniform global observations of meteorological parameters for the first time.

And if computer power gives us remarkable new capabilities for observing the atmosphere, it is even more important for interpreting those observations to model and predict the weather. The atmosphere is one of the most complicated systems that humanity has ever made a sustained effort to understand, model, and predict. And not only is it unbelievably complex, but it is also inherently nonlinear. All of fluid mechanics is fundamentally grounded in a set of nonlinear equations, the Navier-Stokes equations, that were discovered in the nineteenth century and whose solution has bedeviled mathematicians, engineers, and meteorologists ever since. The author of the Web site for the Navier-Stokes equations writes that "according to the database Web of Science (Science Citation Index) there is an average of 15–20 published papers per week dedicated to the subject." The Web site lists 11 scientific conferences related to the subject in the period from March to June of 2002 alone. Nonlinear problems are like that—they are often ridiculously simple to state (the Navier-Stokes equations can be written in one or two lines of algebraic and calculus notation), and they are impossibly difficult to solve without computers, except for special cases with high degrees of symmetry. But with massive amounts of computer power they become merely very difficult to solve.

Surprisingly, the first serious attempt to perform a numerical weather prediction predates the computer era by several decades. One eccentric mathematician, working in impossible conditions during World War I, made the first serious attempt to model the atmosphere numerically. Brian Hayes describes the effort (2001, p. 10):

> In the winter of 1917 Lewis Fry Richardson was driving an ambulance for a French infantry division on the Western Front. At the same time, in spare moments behind the lines, he was completing a vast project of mathematical calculations. He had set out to compute the weather, to predict the temperature, the winds and the barometric pressure from

> first principles of physics. It would have been an ambitious undertaking in the best of circumstances—in a quiet study with a long oak table for spreading out the paperwork. Richardson did it in the middle of a war that was, among its other distinctions, the muddiest in human history. "My office," he wrote, "was a heap of hay in a cold rest billet."
>
> Richardson's goal was to follow the development of the weather for a six-hour interval in a small area of central Europe. Even a forecast of this limited scope called for a calculation of daunting complexity. Needless to say, he had no electronic computer to do the arithmetic for him. He worked with pencil and paper, as well as a slide rule and a table of logarithms. To guard against careless mistakes he did everything twice.
>
> The outcome was not a conspicuous success. The actual weather changed little over the period of the forecast, but Richardson's equations had the barometer rising fast enough to make your ears pop. The calculation was a brilliant piece of work nonetheless. Although he failed to predict the weather, he predicted the future of weather prediction. The forecast you now see on the six o'clock news is based on simulations remarkably like Richardson's (including the occasional error).

Richardson used data that were averaged over 25 sections or blocks of atmosphere that were about 200 kilometers across. These blocks were divided vertically into cells at elevations of roughly 2, 4, 7, and 12 kilometers. Thus, his calculations covered 125 cells. Richardson used 23 of his 25 blocks as input data and then actually calculated the weather changes for only two of the central blocks. And he only calculated the change in the weather over two three-hour intervals. This is extremely limited, of course, and it wasn't even a prediction of the weather. Richardson's data were several years old. Using the technology available in 1917, it was not possible to collect data and perform the massive calculations in time to actually forecast the weather in the future.

Richardson went on to propose an incredible computer scheme, using 64,000 parallel-processing calculators to automate his procedures and calculate weather patterns for the entire globe. His proposed 64,000 calculators were all human, of course, men and women with slide rules and logarithm tables. Electronic computers were about three decades in the future.

Today very similar calculations are done with massive amounts of electronic computer power. The computer models that let us "see" the weather and observe its behavior are staggering in their complexity. For example, the European Center for Medium-Range Weather Forecasts (ECMWF) runs a model that calculates weather parameters on a grid, whose average spacing throughout the global atmosphere is about 40 kilometers. This may seem extremely coarse, but it means that the model is keeping track of some 20 million grid points around the globe. And each grid point has to model more than just temperature, barometric pressure, and relative humidity. There are many other relevant parameters that the computers have to keep track of at each grid point, including sun location (angle above the horizon), cloud cover (which affects the amount of sunlight at the surface during the day and thermal radiation at night), water/ice fraction in clouds, wind speed and direction, chemistry, including ozone and carbon dioxide concentration, and so forth. The models also have to keep track of parameters that model the behavior and effects of the Earth's surface underneath the atmosphere, including water temperature over oceans, soil temperature and moisture over land, vegetation, terrain height, snow cover, and depth.

These models assimilate about 200,000 input observations every six hours, including classical observations of temperature, pressure, humidity, and wind speed and direction from ground stations, plus weather radar data, reports from aircraft and ships, and satellite remote sensing data. Yet even these models are woefully inadequate. A great deal of weather can take place within the 40-kilometer scale of the cells of these models, and the models are too coarse to see it. Also, the input data for the models are imperfect. Even satellites cannot be everywhere at once, and their data have uncertainties and problems of interpretation of the remote sensing data.

The parameters for the ECMWF models are updated at 20-minute intervals throughout the day. Needless to say, no reasonable amount of effort in the absence of modern electronic computer power could make these calculations at all, let alone make them fast enough to be useful for forecasting the weather. There is a running joke in the meteorological community about a computer program that could produce a perfect prediction of tomorrow's weather, but unfortunately the program takes 20 years to run (on current computers). The calculations that assimilate the weather data available today and extrapolate the weather patterns for the next few days are complex enough to push the limits of the most powerful computers that are available. The computer just delivered to the ECMWF is one of the most powerful ever made for the civilian community. It has the capacity to run seven teraflops, or seven trillion floating-point arithmetic operations per second, and is scheduled to be upgraded to some 20 teraflops. (Of course, any discussion of record-breaking computer capabilities today is likely to appear amusing, even silly, from a standpoint of a few years into the future when computer technologies will outstrip our present capabilities by large numbers of orders of magnitude. Critics from the future should be aware that we are doing the best we can today, and that they themselves are subject to the same difficulty and criticism from their own futures.)

The ECMWF Web site at http://www.ecmwf.int/services/data/technical/index.html can be consulted for more details about these computer models and data assimilation levels. The Web site has a significant advantage over this book in that it can be routinely updated as capabilities change, so it is not in danger of becoming obsolete.

These numbers from the ECMWF Web site give us several different ways of calculating the magnitude of the phase change associated with the application of computer technology to problems in meteorology. We could begin with just the simple problem of collecting data. The first attempt in the precomputer era to collect weather data from a broad geographic region was the Smithsonian's efforts using telegraphic dissemination of the data. According to Monmonier (1999, p. 40): "The [Smithsonian Institution] report for 1857 describes a clerical staff of "'twelve to fifteen persons, many of them female' inundated by 'the records of upwards of half a million separate observations [per year], each requiring a reduction involving an arithmetical calculation.'" In contrast, as noted above, the ECMWF handles 200,000 input observations every six hours, or about 300 million observations per year, an increase of a little less than three orders of magnitude.

The phase change involved in the numerical modeling problem is even more dramatic. Richardson's proposed 64,000-person computer was an increase of about five orders of magnitude over his own heroic efforts, but it was never implemented, of course. It would have been expensive and error-prone, and it is dwarfed by modern computers. If we assume that each of Richardson's human computers could multiply two numbers on their slide rules in one second (and ignore the difference between slide-rule accuracy and the accuracy of the standard 64-bit or 16-decimal-place computer calculations), then this massive effort would have amounted to 64,000 flops (floating-point operations per second) or .064 megaflops. In contrast, the supercomputer at the ECMWF performs 7,000,000 megaflops, an increase of an additional eight orders of magnitude over Richardson's impractical computer proposal or thirteen orders of magnitude over Richardson's actual single-person efforts. Thirteen orders of magnitude is one of the largest phase changes that I am aware of in any field of science, dwarfing even the massive phase change produced by computer technology in astronomy.

Meteorology thus exemplifies the incredible power of the application of modern electronic computers to scientific and technical problems facing modern civilization. Meteorology encompasses a set of problems whose difficulty and complexity defy solution without computers, and yet it is a set of problems that is of vital interest to everyone. Predicting the weather is important not only for those who want pleasant weather for their next picnic or deep snow for their next ski holiday, but it is also of central importance to farmers and the entire food production industry. And it is a life-or-death matter for those whose lives, homes, and livelihoods are threatened by hurricanes, tornadoes, floods, and drought. Computer technology gives us the capability to deal with these problems at a deeper and more effective level than was ever possible in any civilization in the precomputer era.

8

AN INFORMATION-THEORETIC PERSPECTIVE ON PHASE CHANGES AND PARADIGM SHIFTS

This chapter will exploit a set of novel mathematical ideas that will enable us to begin to develop a unified theoretical framework that can help us understand both phase changes and paradigm shifts. This new philosophical framework spans all of the sciences and mathematics and can even extend to broader fields of human knowledge in the humanities. The critical ideas are rooted in the modern mathematics of information theory and, particularly, the framework that Gregory Chaitin calls "Algorithmic Information Theory" or AIT. This application of information theory to the philosophy of science gives us yet another example of the impact of computer technology on the rest of science and mathematics. The fundamental theoretical concepts that stand behind the computer revolution are also important for developing a deeper understanding of phase changes and paradigm shifts, and indeed, for developing a novel approach to the philosophy of science.

Information theory is of central importance to the development of any philosophy of science because it is the one area of mathematics that touches every field of human knowledge. Other branches of mathematics such as arithmetic and geometry clearly have very broad applications, but information theory touches absolutely everything. The reason is simple and obvious: Every branch of human knowledge must, by definition, deal with information in some form. Information theory is therefore germane to every field of science as well as the humanities. No other branch of mathematics can make so sweeping a claim.

Moreover, information theory is relatively new, dating back only to the middle decades of the twentieth century. It provides us with powerful new theoretical tools and fundamental insights that were simply not available to anyone attempting to develop a philosophy of science in any earlier period. It is so new that, to my knowledge, no one has yet tried to develop a philosophy of science that is grounded in the ideas of modern information theory. And yet, as noted above, information theory will be found to be important to every field of study. It is the one tool that is absolutely indispensable to a philosophy of science. Many of the problems associated with attempts to develop a philosophy of science in the past can be traced directly to an inadequate understanding of information theory.

To develop an information-theoretic perspective on the philosophy of science we need to begin by defining some basic terms and fundamental concepts, starting with the concept of information itself. A broad definition of information is that it consists of anything that can be recorded in any form, including text, pictures, music, speech, and so forth. Information is generally recorded in two different forms, analog and digital. Analog recording techniques are perhaps most familiar in old-style vinyl LP record albums, while digital recording is used by more recent compact disk technology. Roughly speaking, analog techniques for recording information in the form of sound or music convert the information into an oscillating line, typically a wavy groove in a plastic record. Digital recording converts the same information instead into a sequence of numbers that represent or encode that information. In modern computerized applications the numbers are generally recorded as binary (base-2) numbers that consist of sequences of 0's and 1's.

The two different techniques for recording information are not as different as they may at first appear. They are completely interconvertible. Electrical engineers have long known that when information is recorded in analog form there are standard information-theoretic techniques for converting the recording to digital form. In other words, any analog recording can be converted to a string of binary numbers (and vice versa) without loss of information, as Nyquist famously proved[1] (see the discussion in Press et al., 1992, pp. 494–495). And if information is already recorded digitally, then it is already in the form of a string of numbers.

The 0's and 1's in these binary numbers are referred to by computer aficionados as "bits." But what exactly do we mean by the word "bit?"

It is a concept that is central to modern information theory and, like many pivotal discoveries, it is an idea that is so simple that it seems obvious rather than revolutionary. Claude Shannon at Bell Labs recognized that the fundamental unit of information measurement is the quantity of information needed to decide between two alternatives (0 or 1, true or false, black or white, and so forth). Shannon coined the word "bit" (from BInary digiT) to refer to this fundamental unit of information measurement. He went on to show that if one bit can decide between two alternatives, then two bits can decide among four alternatives by first dividing the four alternatives into two sets of two alternatives each. One of the bits then decides between the two sets and the other one between the two alternatives within the selected set. Similarly, three bits can decide among eight (2^3) alternatives and so forth. Each additional bit doubles the number of choices that can be made. If a string of bits is N bits long, then the string can distinguish between 2^N alternatives. If we have a 10-question true-false test, for example, then the answer to each question consists of one bit (true or false), and ten bits of information are needed to answer all the questions on the test. There are 2^{10} (= 1,024) possible combinations of true and false answers, only one of which is completely correct. Thus, ten bits of information are needed to choose from among 2^{10} possibilities. As noted above, computers normally represent bitstrings as sequences of 0's and 1's.

Similarly, the information contained in an ordinary page of text can be converted to a string of bits using standard computer codes such as the ASCII code that specifies binary values for letters and punctuation marks. And information contained in visual images such as photographs can be converted to binary numbers by breaking the picture down into small elements called "pixels" (short for "picture elements") and assigning a number that represents a gray-scale value (for black and white images) or a sequence of numbers for different colors in a color image.

We therefore arrive at the fundamental idea that information consists of strings of bits, and, naively, information can be quantified by counting the number of bits in a bitstring. But although the idea of quantifying information by counting bits in a bitstring is simple and conceptually useful, it has a serious defect that Chaitin corrected when he developed AIT; indeed, addressing this defect was one of the principal motivations for the creation of AIT. The problem is that not all bitstrings of a given length contain the same quantity of information.

For example, a bitstring that is determined by a random process, such as a coin-toss, contains a lot more information than a bitstring of the same length that consists of all zeros (see the discussion in Robertson, 1999). (The bitstring that results from a sequence of coin-tosses may not contain very useful information, but it is information nevertheless.)

AIT deals with this problem by developing the concept of compression of information. In AIT the information content of a bit string is defined to be the number of bits in the smallest computer program that will generate that bitstring.[2] Any bitstring that can be generated by a computer program that is shorter than the string itself is said to be compressible. This profound and deceptively simple concept allowed Chaitin to develop a set of remarkable theorems that establish a set of fundamental ideas about compressed information. These theorems can be exploited to help us develop a deeper understanding of both phase changes and paradigm shifts. They also provide critical new perspectives and vital insights that clarify the meaning and significance of the pioneering work of Kurt Goedel and Alan Turing in the middle decades of the twentieth century.

One of the most important results that AIT is able to establish concerning compressed information is that it is governed by a conservation law, very similar to conservation of energy in physics[3] (Chaitin, 1999, p. 108). Just as the quantity of energy is unchanged under physical operations (i.e., forces, collisions), the quantity of compressed information is unchanged under logical operations. In physics, the law of conservation of energy implies that the quantity of energy output from any machine must be less than or equal to the quantity that was input. Similarly, in AIT, the conservation of compressed information implies that the quantity of information output from any computer program (and, equivalently, from any sequence of operations in formal logic) must be less than or equal to the quantity that was input. Or, as Chaitin succinctly put it, one of the key goals of AIT is to show that "if one has ten pounds of axioms and a twenty-pound theorem, then that theorem cannot be derived from those axioms" (Chaitin, 1990, p. 62). Using Chaitin's definition of compressed information, this becomes almost a tautology: any bitstring that is generated by a program that is smaller than the bitstring is compressible by definition.

The concept of conservation of information under logical operations places some interesting theoretical limits on the capabilities of conventional computer programs. It establishes, for example, the impossibility of creating a computer program that would generate or discover a new mathematical axiom, an axiom that was not built into the program in the first place or fed to the program as input data. Generating a new axiom would violate conservation of information. Again the argument is simple: All computer operations are equivalent to operations in formal logic, and no logical operation can generate any statement that is logically independent of the axioms that it started with. And axioms, by definition, are logically independent of each other. As pointed out in Robertson (1999, p. 32), both Alan Turing and Douglas Hofstadter made an error on this point by assuming that a computer program could be written that would generate axioms. And Barrow makes a similar error (1998, p. 222) when he says that "Condition 1 [that mathematics must be finitely specified] means that there is a listable infinity of axioms. There must be a definite algorithmic procedure for listing them." It is clear from simple considerations of conservation of information that there can be no finite algorithm that would "list" an infinity of axioms. Axioms are the one thing that cannot be generated by logical operations on other axioms.

Chaitin was able to establish some further surprising results about compression of information, as noted in (Robertson, 1999):

> AIT is able to establish a number of fundamental results about the compressibility of bit-strings. For example, it might seem, naively, that any bit-string could be compressed if one were simply clever enough to invent the necessary algorithm and write the corresponding computer program. But this is not the case. One of the fundamental results from AIT is a proof that nearly all bit strings cannot be compressed. They cannot be "computed" using fewer bits than the string itself (Chaitin, 1990, p. 15). This theorem is closely related to a theorem by Alan Turing that showed that nearly all numbers cannot be computed by a computer program (the "uncomputable" numbers; see Robertson, 1998, chapter 4).
>
> Another strange result from AIT is that in most cases it is not possible to determine whether a bit-string is compressible. If it happens that a bit-string is compressible, then that fact could be demonstrated by constructing the necessary compressing algorithm, the computer program that will generate the string. However, if a bit-string is not compressible (as nearly all bit strings are not), then there is no way to prove this fact, no way to determine that a compressing algorithm does not exist. (See Chaitin, 1990, pp. 30–32)

The concept of compressibility of information is one of the central ideas that can help us understand the fundamental operations that are the basis of both the sciences and the humanities. Following up on some ideas that were first discussed by Chaitin (1970) and Solomonoff (1964a and 1964b), we can start with the simple observation that every field of human knowledge is based on the collection and compression of information. Furthermore, one of the key differences between the various branches centers on the amount of compression that can be attained.

In order to define both the sciences and much of the humanities in terms of collection and compression of information, some clarification and amplification of these concepts is needed. For example, the notion of collecting information involves more than just recording or writing down bitstrings. In order to be useful the process of collecting information must include both verification and evaluation of information. Verification of information involves the essential concept that experiments and observations should be repeatable and should, in fact, be repeated under varied and carefully controlled conditions. Indeed, the concept of repeatability of experiments, observations, and measurements has long been recognized as central to the process of collecting information in both the sciences and the humanities. In history, for example, the need to find two or more primary sources that attest to a postulated historical fact is widely recognized. And because information is almost never free of errors, the process of evaluating information involves the often difficult procedure of measuring or determining quantitative bounds on the error levels of the information. The concept of "collection of information" therefore covers both the experimental part of the sciences and the observational processes that are found in all fields of knowledge and is closely related to the concept of a phase change as defined in chapter 1. A phase change generally occurs following the invention of a novel technology for collecting information.

If the collection of information subsumes the experimental and observational portions of intellectual activity then the second part of the definition, compression of information, covers the theoretical part and is closely connected to the concept of a paradigm shift. The word "compression" as used in this definition is intended to cover both compression and decompression of information. To compress information we start with a large quantity of raw observational and experimental information and reduce it to a theory that is based on a small number of fundamental principles and ideas (a paradigm). In principle, if the data that have been compressed are expressed in bitstrings, then the theory or paradigm can be expressed as a computer program that will generate those bitstrings or will generate a suitable approximation that is within the known errors of the data being compressed.

The process of decompression is then simply a matter of running the computer program to generate or represent the original data set. Further decompression can then be done to predict new observations that might be made that lie beyond the domain or range of the original set of data. Comparisons of these theoretical predictions with new observations, measurements, or experimental results can then be used to test the limits of a particular form of compression (theory) and verify both its accuracy and the range over which it is useful. This comparison of decompressed theories with newly collected information can also be used to reject a particular theory or model if the magnitude of the disagreement with observations is found to be inconsistent with the estimated errors in those observations. The concept of compression of information therefore gives us something that Thomas Kuhn lacked—a precise definition of the term "paradigm." A paradigm could be defined as a method or set of algorithms for compressing information.

All of the theoretical side of science can be considered (or modeled) as a set of attempts to create short and efficient computer programs that will generate particular bitstrings. The desired bitstring might describe, for example, the positions and motions of planets and satellites, the results of a particle accelerator experiment, or the quantity of gold or oil that can be found in a particular patch of ground.

To take a simple example, one of the earliest examples of compression of information is found in Ptolemy's use of epicycles to calculate the positions of planets. Ptolemy began with an enormous quantity of information about planetary positions that were largely taken from Babylonian records. He compressed that information to a set of parameters for the epicycles that matched each planet's motion. Ptolemy's theory not only accounted for the observed motions of planets in the past, it also had the signal virtue that it could be decompressed to predict planetary positions in the future with reasonable accuracy levels.

Ptolemy's theory was supplanted by that of Copernicus, which achieved a significant additional compression of information. Copernicus knew that many of the epicycles needed for the different planets in Ptolemaic theory are identical in period and related in phase. These particular terms could, in fact, result from the parallax effects of the motion of the Earth rather than from the motions of the other planets.

The Ptolemaic epicycles could therefore be replaced by uniform circular motions of all the planets including the Earth. The Copernican theory that resulted was simpler and more "compressed" than Ptolemy's because uniform circular motion is much simpler (and easier to calculate) than epicyclic motion (although in the end Copernicus was forced to add complexity to his model to try to compensate for the fact that uniform circular motion is not a sufficiently accurate model for the motions of the planets).

Tycho Brahe then performed one of the earliest recorded examples of decompressing the information from two different mathematical theories, Ptolemy's and Copernicus's planetary theories, and comparing both of them with an observation of a planetary conjunction (Eisenstein, 1979,vol. 2, p. 624). When he found that neither of them agreed with his observation very well, he built an observatory in order to collect more information. Brahe's observatory was a brilliant success: His data were the most important observations made in astronomy prior to the invention of the telescope.

Kepler's work with Brahe's planetary data was an early example of the use of observed information to reject a mathematical theory, that is, to show that a particular compression of data was inadequate to match the error tolerances of that data. Kepler was able to show conclusively that the assumption of uniform motion along circular orbits could not be made to fit Brahe's superb new data for the motion of Mars in particular. The discrepancy to the Copernican model was small, the famous "eight minutes of arc," but Kepler knew that it was larger than the errors in Brahe's data, which were about one minute of arc. He then showed that motion along elliptical orbits together with his "equal area in equal time" rule would fit Brahe's data to within its error tolerances. Kepler's theory was *less* compressed than Copernicus's, partly because ellipses require more information than circles, but its increase in accuracy more than compensated for its increase in complexity.

Newton's theory of gravity and dynamics then achieved one of the greatest compressions of information in all history. Newton's theory was significantly more compressed than Kepler's because it was based on only three laws of motion plus the inverse-square law of gravity. And it was not only more simple, it was also far more accurate. It predicted such things as planetary perturbations of the Keplerian ellipses that had not yet been observed. Even more important, it extended to far more phenomena than just planetary positions. It explained tides, precession of the equinoxes, the equatorial bulge of the Earth, and the dynamics of terrestrial as well as celestial bodies to an accuracy level that would not be surpassed for centuries. Indeed, the need for an improvement on Newtonian theory would not be suspected until Leverrier discovered the anomalous perihelion precession of Mercury in the nineteenth century.

Clearly then there is frequently more than one way to compress a given set of data (bitstring). All compressions are not created equal, and there are often trade-offs involved in deciding between various compressions. If the accuracy of two different compression schemes is identical, then the one that caused more compression, the smaller program, is generally preferred. But in the case of the replacement of the Copernican model of uniform circular motion with the Keplerian model of nonuniform elliptical motion, the trade-off was one of increased complexity for increased accuracy. In this common trade-off, researchers often use a less complex model where high accuracy is not required. For example, astronomers today frequently use simple Keplerian orbit models when the higher accuracy of Newtonian or relativistic mechanics is not required.

There are also cases where apparently distinct physical theories are mathematically identical; in perhaps the most familiar case, Werner Heisenberg's matrix formulation of quantum mechanics and Erwin Schrodinger's wave-equation formulation were shown to be mathematically identical (see the discussion in Gray, 2000, p. 161). But even in cases where two different compressions are mathematically identical, there is often reason to favor one over another. For example, the sequence of digits in π can be compressed with a number of different formulae, such as the following infinite series and continued fraction:

$$\pi = 4\left(1 - \frac{1}{3} + \frac{1}{5} - \frac{1}{7} + \frac{1}{9} - \dots\right) = 3 + \cfrac{1^2}{6 + \cfrac{3^2}{6 + \cfrac{5^2}{6 + \cfrac{7^2}{6 + \cfrac{9^2}{6 + \dots}}}}}$$

Although both expressions give the same value when carried to enough terms, the continued fraction converges much faster than the series. For example, after 100 terms the error in the continued fraction is about 0.2×10^{-6}, and the series does not achieve this accuracy until more than 4,000,000 terms are added. Thus, even though these two compressions are mathematically identical in the limit as the number of terms goes to infinity, one of them is still much more useful than the other (and neither of them converges as fast as the formula given in chapter 5, which achieves even better accuracy when only the first four terms are added).

This definition of human knowledge in terms of collection and compression of information appears to model much of what is actually done in research activities. But if the definition is going to be useful, it should provide novel insights into those activities, as I believe it will. The critical insights are related to five simple facts (see Chaitin, 1975; 1990):

1. Compressed information is conserved under logical operations.
2. Nearly all bitstrings cannot be compressed.
3. If a bitstring cannot be compressed, there is no algorithm that will demonstrate that fact.
4. The quantity of compressed information (the number of axioms) needed for mathematics is not finite (Chaitin's powerful and insightful extension of Goedel's theorem).
5. There is no finite algorithm that can be applied to every mathematical proposition that is guaranteed to prove the proposition either true or false (Turing's theorem on the Halting Problem).

The first three of these facts are essential results from Chaitin's AIT, as discussed above. The fourth and fifth are discussed in Robertson (1998, pp. 37–56). Briefly, in a famous address in 1928, David Hilbert asked for a proof that mathematics is complete, so that every proposition could be proved either true or false. In 1931, Kurt Goedel's celebrated incompleteness theorem succeeded in proving the converse, that mathematics is essentially incomplete. In other words, for any finite number of axioms (beyond some minimum number[4]) there is always an infinite number of true statements that cannot be proved. Therefore, the number of axioms or quantity of compressed information that is needed to complete all of mathematics cannot be finite. In the same address, Hilbert asked for a proof that mathematics is decidable, in other words that there exists a [finite] method or algorithm that can be applied to any mathematical proposition that is guaranteed to prove the proposition either true or false. Alan Turing succeeded in showing that the very existence of such an algorithm would lead to a contradiction, and that therefore no such algorithm exists.

All five of these facts have important implications for any philosophy of science. We have already examined some of the consequences of fact number 1 above. The second of these facts tells us that most fields of human knowledge must deal largely with information that cannot be compressed, simply because the preponderance of all information must be in this form. However, from fact number three, we cannot generally tell what information can be compressed and what cannot. We should expect, then, that various disciplines will differ widely in the amount of compression they are able to attain. We might try to classify various disciplines according to the degree of compression that they employ. For example, mathematics uses information in its most compressed form (axioms). All the information that is contained in all the fields of mathematics that are presently known can be compressed into a few tens of axioms. This form of information is so compressed that mathematicians spend nearly all of their time decompressing it. The action of developing conjectures and proving them as theorems is the essence of decompression. This is not to say that mathematicians do not collect information. On the contrary, as Stewart observed, great mathematicians have frequently done experimental calculations to develop insights in preparation for the development of more general results and proofs (1992, pp. 314–315). These experimental calculations are a form of collection of information. But compression and decompression of information are carried farther in mathematics than in any other field.

Physicists employ collection and compression in roughly equal quantities. The experimental and theoretical portions of physics are generally recognized as coequal partners in the enterprise (although individual researchers frequently exhibit a preference for one over the other). But as we move away from physics and mathematics, compression of information rapidly becomes much more difficult. For example, geology and biology each have their own overarching theories (compressed forms of information), including such theories as plate tectonics and Darwinian evolution. Although these theories have succeeded in compressing a great deal of raw data, the degree of compression is less than in physics and decompression is far more difficult. As Stephen Gould frequently observed, you cannot use Darwinian theory to predict the future course of evolution or even to explain, for example, why the dominance of vertebrates that is observed in present-day ecosystems evolved from the populations of organisms that existed in the Cambrian.[5] Plate tectonics similarly cannot predict the future development of continents and oceans beyond a geologically brief time interval.

Moving from the hard sciences into the social sciences and humanities, we enter fields where compression becomes even more difficult and decompression is nearly impossible. In the study of history, for example, the collection, verification, and evaluation of information is as important as in any other field. Yet here compression of information has been found to be, at the very least, problematic. Of course, historians have frequently tried to formulate "laws" of history (familiar examples include the works of Marx and Toynbee), but such theories do not find wide acceptance today.

Historians and others in the humanities are therefore accustomed to dealing with the problems of handling vast quantities of information that are fundamentally incompressible, and they have therefore been forced to develop techniques for dealing with such vast quantities of information. The basic difficulty, of course, is that human minds are finite—there is a limit to the quantity of information that one individual (or even groups of individuals) can learn and manipulate. This is one of the reasons that compression of information is so important (besides the fact that the compression algorithm is often intrinsically interesting): Enormous quantities of information can be reduced to comprehensible patterns, compressed expressions of the same information. But the only option that remains when the sheer quantity of incompressible information exceeds our ability to comprehend it is some form of selection. We must necessarily ignore most of the available information and focus only on the areas that are of maximum importance to us. As Grafton observed: "without oblivion, history could not continue to be written" (1997, p. 230).

Of course, the process of selection of information cannot be simply random. When compression fails and selection must be used, organization takes the place of compression. Information is collected, organized, and indexed so that vital information can be found when it is needed and ignored the rest of the time. In history, for example, a variety of classification schemes have been developed to try to make sure that the relevant information relating to one historical period or personage can be located. Thus, historians define fields such as "Modern American History," and "France in the Age of Louis XIV," information fields that are classified and restricted in both time and space.

Of course, no classification scheme is perfect, but such schemes along with indexing and other search techniques are the only known ways to handle vast quantities of irreducible information.

There are great differences between various disciplines on the question of the utility and "naturalness" of classification schemes. Some areas, such as taxonomy in biology, have very natural classification schemes based on the existence and relationships of biological species. Other areas such as history and archaeology have classification schemes that are much more arbitrary, for example, divisions based on time intervals or geographic regions that may reflect little more than the interests of the particular researchers. And even in taxonomy there are serious difficulties involved in organization above the species level, where terms such as "genus," "family," and "order" are widely recognized to be arbitrary. Despite these difficulties, classification schemes are an essential element of the process of research.

Therefore, to finish the definition that would attempt to unify all the sciences and the humanities, we need to add a third component. All areas of intellectual activity involve the collection, compression, and organization of information. And compression and organization are complementary. Organization takes over where compression fails, as it must at some point because nearly all information is incompressible. And computer technology gives us vast new powers to organize information (as well as collect and compress it), through the use of massive computer capabilities for the storage of vast amounts of data, combined with powerful software and novel algorithms that allow us to search and sort through such enormous quantities of information. Various World-Wide-Web search engines represent some of the first and probably primitive examples of such software.

This information-theoretic viewpoint on science and mathematics can provide insights into one of the important questions about phase changes in science and mathematics that was raised in chapter 1. Since a phase change is usually preceded by some critical state of a system, it might be useful to develop a conjecture about the nature of the critical state that precedes a phase change in science and mathematics. Of course, science and mathematics are very complex systems and it is quite possible that there are a number of different ways that they can approach critical states. But if we think of research as consisting of the collection, compression, and organization of information, then the process begins with the collection of information, with observations and experiments such as Brahe's measurements of planetary positions or Rutherford's observations of the scattering of alpha particle radiation or Mendel's observations of pea-plant inheritance. Theorists then work to compress that newly collected information. A critical state can occur because there is only a finite number of ways to compress a given quantity of information. And in practice the number of useful compression schemes is not only finite, it is frequently quite small. For example, in two thousand years we have only devised a small number of general schemes for compressing planetary position data: Ptolemaic epicycles, Copernican circular motions, Keplerian ellipses, Newtonian dynamics, and Einsteinian general relativity.

Thus, research fields can grow stale as the various possibilities for compressing a given set of data or a given type of observing technique are exhausted. At that point a critical state is reached, where further progress is stymied until some novel information from newly developed observing schemes becomes available. One of the classic examples of such a need for new information is found in astronomy prior to Galileo's development of the telescope. Up to this point, observational astronomy had consisted exclusively of naked-eye observations. And naked-eye observations had produced a wealth of information about the universe, about the locations of the fixed stars, the precession of the equinoxes, and finally, perhaps the crowning glory of the era of naked-eye observations, Kepler's three laws of planetary motion. But with Kepler's work and Newton's use of Kepler's laws in the development of Newtonian mechanics, the possibilities for naked-eye astronomy were essentially played out. I know of only a few examples of work that depended on naked-eye observations in astronomy since the time of Kepler and Newton (including such things as the recovery of ancient records of eclipses of the sun). But the possibilities for novel information collection and compression schemes were essentially exhausted after Kepler's day, and the field was ripe for novel observing techniques that were typified by Galileo's telescopic observations.

This description is oversimplified, of course. There are cases where a research field appears to have stagnated and be in need of new information when what is needed is simply a new insight into extant data (a novel compression algorithm). But even in these cases the field again often approaches a state in which novel forms of information are needed for further progress. Thus, one possible critical state that precedes at least some of the phase changes in science and mathematics is a state in which the possibilities for the compression of a given set of information have been exhausted, as naked-eye astronomy was essentially exhausted following Kepler and Newton.

One of the central problems for a philosophy of science is the question of whether this process of collecting and compressing information will be an infinitely recurring process. Will researchers forever be exhausting various forms of data and then turning to new techniques for observing and discovering new forms of data that transform the field once again, so that the process repeats indefinitely, or will the process terminate at some point?

This question is related to one of the classic problems in the philosophy of science, as stated by Kitcher (1993, p. 6): "Can we legitimately view truth as a goal of science?" There are two possible answers to this question: Either there is an ultimate truth that can be sought with scientific methods or there is not. And in the first case, if there is an ultimate truth that can be sought, then there is another vital question that is raised by the concept of compressible information from AIT: Can that ultimate truth be expressed with finite quantities of information or not? Finally there is a closely related question about the "scientific method," the fundamental techniques that are needed to attain, approach, or approximate that ultimate truth: Can the scientific method be expressed with finite quantities of information or not? Before the development of modern information theory we lacked the terminology and basic concepts that are needed to frame these questions quantitatively. Yet they are of central importance to the philosophy of science because the very nature of philosophy itself depends on the various possible answers to these questions.

Before I examine these questions I need to clarify what I mean by the phrase "ultimate truth" in the last paragraph. I am referring to the fundamental principles (the incompressible information) that physics is based on. I do not mean to include the "historical accidents" of physics such as the masses and orbital parameters of the planets as part of this "ultimate truth" (although these are matters of intrinsic interest in many branches of science). In other words I am assuming that physics can be axiomatized like mathematics, and I am asking whether the total number of axioms needed is finite or not. As noted above, in mathematics we know that the total number of axioms needed is not finite. With this definition of the term "ultimate truth" it is clear that the search for this truth is only one component, albeit an interesting and important component, of the overall process of collecting, compressing, and organizing information.

We can therefore consider the answers to the questions above in terms of three possibilities, three possible distinct and different types of universe, each of which is classified according to its underlying information requirements. In a type 1 universe, both the absolute, ultimate truth (the axioms of physics) and the methods that are needed to discover that absolute truth ("scientific methods") can be contained within or expressed with a finite quantity of information. In a type 2 universe the axioms of physics exist but they cannot be contained within any finite quantity of information, and neither can the methods needed to develop successive approximations to that absolute truth. A type 3 universe would be perfectly chaotic: It would contain no absolute truth (there would be no underlying axioms of physics or at least none that are independent of social context, as the postmodernists would have us believe; see the discussion below), and there would be no method that can attain or even approximate any absolute truth.[6]

And as we noted above, the philosophy of science would be a radically different thing in each of these three types of universe. In a type 1 universe, the philosophy of science would be straightforward—it would consist of developing and using the finite methods needed to attain the finite absolute truth. In a type 2 universe, the philosophy of science is a more difficult but not completely intractable problem that we will try to deal with more fully below. In a type 3 universe no philosophy of science is possible; indeed, no science at all is possible.

Prior to the invention of AIT, no one had access to the basic information-theoretic concepts that are needed to frame philosophical questions quantitatively in terms of these three possibilities. Many early attempts to develop a philosophy of science ran into serious difficulties that stem directly from a lack of awareness of them. A type 1 universe was implicitly assumed in essentially all the formulations of a philosophy of science through at least the early decades of the twentieth century, and the assumption remains popular today. Indeed, prior to the work of Goedel, Turing, and Chaitin that established the impossibility of a type 1 universe in mathematics in the mid- to late-twentieth century, no one seriously considered any other possibility. At the close of the nineteenth century, physicists almost universally believed that not only were the fundamental laws of physics finite in number, but they also believed that all of them were already known. No less a personage than America's first Nobel Prize winner, Albert Michelson, famously commented that the only thing remaining to do in physics was to add a few decimal places to the facts that were already known.[7] But the revolutions of the early twentieth century, centered on quantum theory and relativity, shook the confidence of both scientists and philosophers.

A number of philosophers in the early twentieth century developed a set of ideas that we might broadly label "logical positivist" partly in response to the shock of these developments in physics (although the question of exactly who should and should not be labeled a logical positivist is a terminological problem that is beyond the scope of the discussion here). The positivists tried to base all of philosophy on modern science, and they implicitly accepted the idea that science is an unterminating sequence of approximations to an ultimate truth that cannot be reached with any finite amount of information. Karl Popper famously argued that theories cannot be proved correct, only falsified. The positivists avoided any attempt to define absolute truth in terms of a particular and finitely specified physical theory as had been commonly done in the past, as, for example, Newtonian physics was once thought to be an absolutely true model for the behavior of matter and energy, and Euclidean geometry was thought to be an absolutely true model for physical space. Instead, the positivists attempted to specify the "scientific method," the set of techniques that would be acceptable for use in approaching scientific truth. They believed that this scientific method that they were trying to define could be completely specified using a finite amount of information, and they set out to try to do so. As Chalmers (1990, pp. 3–4) describes the positivist position:

> The key aim of the logical positivists . . . was to defend science. . . . They endeavoured to construct a general definition or characterization of science, including the methods appropriate for its construction and the criteria to be appealed to in its appraisal. With this in hand, they aimed to defend science and challenge pseudo-science by showing how the former conforms to the general characterization and the latter does not. . . . the general strategy involved in the positivists' attempt to defend science is still widely adhered to. That is, it is still commonly assumed, among philosophers, scientists and others that if science is to be defended we require a general account of its methods and standards. . . . The positivists . . . sought a "unified theory of science" (Hanfling, 1981, ch. 6) which they could employ to defend physics.

Chalmers goes on to describe the work of recent positivist philosophers:

> Imre Lakatos and Karl Popper are two prominent philosophers of science in recent times who adopt the positivist strategy . . . Lakatos (1978, pp. 168, 169, and 189) considered the "central problem in phi-

> losophy of science" to be "the problem of stating *universal* conditions under which a theory is scientific". . . . Popper (1972, p. 39; 1961, section 29) himself sought to demarcate science from non-science in terms of a method that he saw as characteristic of all science, including social science. . . . Thus two contemporary physicists (Theocharis and Psimopoulos, 1987) urge that the practice and defence of science should involve an appeal to an adequate definition of scientific method and deplore the extent to which practising scientists are ignorant of such a definition.

At first this positivist position seemed very reasonable, but it ran into serious difficulties when positivists tried and failed to agree on a statement of the universal method of science, as Chalmers described in some detail (1990, pp. 11–23). Such difficulties would be expected and, indeed, would be unavoidable in a type 2 universe where no finite quantity of information would suffice to specify the "scientific method." The positivist's major error lay not so much in the assumption that the scientific method could be specified with finite amounts of information, which might well be correct, but in neglecting or being completely unaware of any other possibility. By failing to acknowledge the possibility of a type 2 universe (indeed, many of these philosophers were working some decades before the concepts needed to express the idea were developed), the positivists left themselves open to serious criticism, notably by the postmodernists, who are perhaps typified by Paul Feyerabend. (Just as with the logical positivists, there is a serious terminological difficulty in deciding who should and should not be described as a postmodernist.)

Feyerabend correctly noted that not only had the positivists failed to construct a complete statement of scientific methods, but also that major advances in the history of science had often involved the use of concepts and "hypotheses that contradict well-confirmed theories and/or well-established experimental results" (1975, p. 29). For example, Copernicus assumed that the Earth moves, in contrast to everyone's ordinary perception, Newton assumed "action at a distance" in his formulation of gravity, and Einstein assumed a non-Euclidean physical space-time in his mathematical model for gravity. Feyerabend (1975, p. 295) then went on to write that "the idea that science can, and should, be run according to [finite] fixed and universal rules is unrealistic and pernicious." This statement is perfectly correct if the physical universe is a type 2 universe. But then he goes to advocate what he calls total anarchy: "All methodologies have their limitations and the only 'rule' that survives is 'anything goes'" (1975, p. 296). In saying this Feyerabend makes a classic logical error of assuming a false dichotomy. Feyerabend assumes that the only alternative to a type 1 universe is a type 3 universe; he is unaware that the possibility of a type 2 universe renders his dichotomy false.

But, as noted above, Feyerabend was not the first to assume this dichotomy. It was first stated by the positivists themselves, who asserted that the only alternative to a complete and finitely specified scientific method was absolute chaos. Chalmers quotes the positivist position as follows (1990, p. 7):

> Either we have absolute standards as specified by a [finite] universal account of science or we have sceptical relativism, and the choice between evolutionary theory and creation science becomes a matter of taste or faith.

and (1990, p. 8):

> Advocates of the positivist strategy typically present themselves as the defenders of science and rationality and their opponents as enemies of science and rationality. In this they are mistaken. In adopting a strategy for defending science that is doomed to failure they play into the hands of the anti-science movement that they fear. They make Paul Feyerabend's job too easy.

Thus, even the possibility that the physical universe might be of type 2 (whether it actually is or not) is extremely useful to a philosophy of science, since it eliminates the dichotomy between a type 1 and a type 3 universe that provided the major and perhaps only serious argument that Feyerabend and other postmodernists could make for the existence of a type 3 universe. The very possibility of a type 2 universe allows us to steer carefully between the Scylla and Charybdis of the positivists and the postmodernists. It allows us to escape the narrow confines of the prison that the positivists would put us in, the straitjacket of requiring a finitely specified scientific method, without necessarily putting us into the nightmare universe of pure chaos that the postmodernists would have us believe, in which astronomy and astrology would be equally ineffective.

I do not mean to imply by the metaphor of Scylla and Charybdis that there is any parity or rough equivalence between the errors of the positivists and the postmodernists. Odysseus understood that Scylla was far less dangerous than Charybdis. Similarly, the positivists' error generally leads to good, if limited, scientific practices. In contrast, I have never been able to find any virtue at all in postmodernist ideas. Much of postmodernist criticism of scientific methods consists of little more than what Gross and Levitt would call "unalloyed twaddle" (See Gross and Levitt, 1994, for a detailed discussion). Postmodernist ideas only occasionally rise to the level of being wrong, but when they do so it can be useful to examine exactly why they are wrong. There is a sense in which Feyerabend's statement "anything goes" is correct. In the development of a theory or paradigm, of a novel means of compressing and decompressing observed information, it is often true that "anything goes." Theorists are free to postulate things that are contrary to accepted views and even contrary to "common sense," as with Copernicus's idea of the Earth in motion and Newton's action at a distance. Similarly, observers and experimenters are free to experiment with novel techniques such as Galileo's telescope and Rutherford's alpha particle beams.

But Feyerabend neglects the next vital step entirely. The theorist must demonstrate that his postulates can indeed be used to compress observed data and compress it better than other ideas, either with greater accuracy or with a broader range of applicability, for example. And the observers and experimenters must demonstrate that their new techniques provide reliable data that can be repeated by skeptical observers under carefully controlled conditions. Novel paradigms and observing techniques that fail these tests can reasonably be rejected. This is essentially a Darwinian "survival of the fittest" process for a given set of theories or compression algorithms and observing techniques. In ignoring this essential feature of scientific methods, Feyerabend is, in effect, proposing a philosophy of science akin to a Darwinian evolution theory that lacks the process of natural selection. And any biologist will tell you that such an omission is fatal, that evolution with only random variation and not natural selection will not work. Neither will a philosophy of science that neglects the essential weeding out of observational techniques that fail to provide reproducible results or of theoretical ideas that fail to produce an improved compression of information.

The possibility of a type 2 universe does more than just destroy the false dichotomy that lies behind much of postmodernist criticism. A type 2 universe is a possibility that many will find intrinsically interesting. It would lie in some sense between the perfectly ordered type 1 universe and the perfectly chaotic type 3 universe. This boundary region between type 1 and type 3 is related to the concept that Stuart Kauffman called the "edge of chaos." Kauffman postulated that this edge of chaos was the only region where life was possible or at least a state that life is found in or evolves toward. This claim leads directly to a further speculation that perhaps life might also be found only in a type 2 universe. This speculation would take us a bit far afield of the concerns of this chapter, but readers who are interested in this idea should perhaps begin with the discussion in (Kauffman 1993).

Chalmers (1990, pp. 20–23) also rejects both the positivist and the postmodernist views and advocates a "third way," based on what he calls "variable methods and standards." The ideas developed here are a variant on Chalmers' ideas, but the concept of compressible information that we now have from AIT allows us to specify ways of testing and validating these "variable methods and standards" empirically, even if those methods and standards themselves cannot be specified completely with any finite quantity of information. To devise these empirical tests, we will need to develop a philosophy of science that is based on concepts and ideas from modern information theory.

The obvious critical question for such a philosophy of science is the nature of the actual universe, its quantitative information requirements. In other words, is the actual universe of type 1, 2, or 3? This is not an easy question, and there is only one area in which the question can be answered with any certainty. AIT, combined with the work of Goedel and Turing, has established that the universe of mathematics cannot be a type 1 universe. Goedel's incompleteness theorem shows that no finite collection of axioms can be complete, and therefore the "ultimate truth" in mathematics cannot be compressed into any finite amount of information. And Turing's halting theorem establishes that there can be no finite method for determining the truth of an arbitrary statement in mathematics and therefore no finite "scientific method" in mathematics either. As the celebrated mathematician John von Neumann put it: "Truth is too powerful to allow anything except approximation" (quoted in Schroder, 1991, p. 371). And as Freeman Dyson noted, Goedel's theorem does not present us with a set of limitations to mathematics but rather with an absence of limits. As Dyson states (1985, pp. 52–53):

> Fifty years ago, Kurt Goedel . . . proved that the world of pure mathematics is inexhaustible. No finite set of axioms and rules of inference can ever encompass the whole of mathematics. Given any finite set of axioms, we can find meaningful mathematical questions which the axioms leave unanswered. This discovery of Goedel came at first as an unwelcome shock to many mathematicians. . . . After the initial shock was over, the mathematicians realized that Goedel's theorem, in denying them the possibility of a universal algorithm to settle all questions, gave them instead a guarantee that mathematics can never die. No matter how far mathematics progresses and no matter how many problems are solved, there will always be, thanks to Goedel, fresh questions to ask and fresh ideas to discover.

But the question remains, is the physical universe a type 1 universe, as many physicists aver (Weinberg, 1992; Lederman, 1993) or a type 2 universe, as Freeman Dyson hopes? Again, citing Dyson (1985, p. 53):

> It is my hope that we may be able to prove the world of physics as inexhaustible as the world of mathematics. Some of our colleagues in particle physics think that they are coming close to a complete understanding of the basic laws of nature. . . . But I hope that the notion of a final statement of the laws of physics will prove as illusory as the notion of a formal decision process for all of mathematics. If it should turn out that the whole of physical reality can be described by a finite set of equations, I would be disappointed. I would feel that the Creator had been uncharacteristically lacking in imagination.

This question was discussed in Robertson (1998, chapter 2, and 2000) and cannot be answered with certainty, although there are some indications that Dyson's hope may be fulfilled, that the physical universe may, in fact, be a type 2 universe. The principal evidence for this (and it is far from being conclusive or even convincing evidence) lies in what Eugene Wigner called the "unreasonable effectiveness" of the applications of mathematics in physics. In other words, mathematics has many applications in physics. If we simply assume that this unreasonable effectiveness will continue through the indefinite future and that mathematicians will continue indefinitely to discover new areas of mathematics that have important applications, then the infinite complexity that we know exists in mathematics would carry over immediately into physics. Furthermore, the failure of the logical positivists to succeed in formulating a complete and finite statement of scientific methods, despite years of effort by very intelligent and knowledgeable specialists, is at least consistent with the idea that the physical universe is a type 2 universe.

On the other hand, if we assume that the physical universe is a type 1 universe, we would then be in the rather odd position of assuming that at some point research in physics is going to end while research in mathematics is still ongoing. In other words, we would be assuming that although progress in mathematics will never end, the development of applications of that mathematics will somehow come to an end. This seems to me to be a very strange assumption, although I know of no way to conclusively prove that it is wrong. As I noted (in Robertson, 2000, p. 26):

> There is a long history of branches of mathematics that were once thought to have no application to physics but were later found to be of central importance. For example, Sir James Jeans once commented that group theory could have no possible application to physics, and Henri Poincare thought the same about non-Euclidean geometry. And today parts of number theory that were once regarded as completely devoid of practical application, especially factoring theory, are found to have practical applications in coding for computer communication. So it appears somewhat perilous to suggest that any part of mathematics is devoid of practical application.

Perhaps all of mathematics has applications, and instead of dividing mathematics between pure and applied branches, it should be divided instead between branches for which applications have already been discovered and branches for which applications have not yet been discovered. If indeed, all of mathematics (or even some nonzero fractional part of it, say half of it or one part in 10^{30}) has applications, then the physical universe would have to be a type 2 universe.

There is one more problem with assuming that the physical universe is a type 1 universe. Even if this assumption is true, there still remain an infinite number of physical problems that cannot be solved. This was also discussed in Robertson (1998, pp. 50–53). Briefly, the problem relates to Turing's theorem again. We begin with one of the simplest possible universes, the one defined by John Conway's celebrated cellular automaton game called Life. In the game of Life the "theory of everything" (TOE) exists, and it can be stated with three simple rules. And given any starting configuration on a Life board, these three rules completely specify the future configurations of the board forever. Yet it has been established that it is possible to construct a Turing machine, a universal computer, on the board of the game of Life. Therefore, by Turing's theorem, there must exist simple questions, such as whether a given configuration will grow without limit, that cannot be answered with any finite algorithm. Therefore, even if the physical universe has a TOE that can be specified with a finite amount of information, if that TOE is sufficiently complex to allow the construction of a Turing machine (a computer) then there is still an effective infinity of problems in physics that cannot be solved, and most of physics would therefore have the characteristics of a type 2 universe even if a simple and finite TOE for physics exists and is completely known.

And even if the physical universe is a type 1 universe, so that it has a TOE that can be specified with a finite amount of information, sciences such as astronomy and biology would still be effectively operating in a type 2 universe, if we define the objective of astronomy to be the study of every star and planet and atom in the universe and biology to be the study of every organism at the molecular level, including studies of the interactions of those organisms and molecules. There is essentially no possibility that problems defined this way can be exhausted in any foreseeable future. And nearly all branches of science have this property, that they cannot be exhausted with any reasonable quantity of compressed information. This is a fundamental consequence of the fact that nearly all information is incompressible.

What about the third possibility: Could the universe be a type 3 universe of complete chaos that Feyerabend and other postmodernists would advocate (although as far as I am aware they never express their ideas in terms of quantitative information requirements)? I am not aware of any way to answer this question with absolute certainty. But although there is no absolute and certain proof, there is some very strong evidence that both the mathematical and the physical universe are not type 3 universes. The evidence is simply the observed fact of the incredible effectiveness of conventional scientific and mathematical methods. In a type 3 universe this would not happen; astronomy and astrology would be equally ineffective, as the postmodernists claim. Although this uncontested effectiveness of conventional scientific techniques does not offer absolute and conclusive proof that the universe is not a type 3, the evidence is far too strong to simply ignore. At the same time there is no particular positive evidence for the existence of a type 3 universe, and the only mildly effective argument the postmodernists were able to offer for it centers on the false dichotomy discussed above.

Thus, mathematics is the only branch of science where we can distinguish with certainty between a type 1 universe and a type 2 universe, and in mathematics a type 1 universe is absolutely ruled out. For the physical universe we cannot absolutely rule out the possibility that it is a type 1 universe, but there are some indications that it might be a type 2 universe, that the TOE for physics cannot be specified with any finite quantity of information. Furthermore, even if there is a physical TOE that can be specified with a finite quantity of information, there must still be an infinity of problems that cannot be solved, and therefore the universe of physics is effectively a type 2 universe. And other branches of science are also operating effectively in a type 2 universe. At the same time, we have clear evidence that both the universe of mathematics and the physical universe are not type 3 universes, where no science or philosophy of science would be possible and mathematics would not be found to be "unreasonably effective."

Suppose we therefore take as a working hypothesis that the physical universe (as well as the mathematical universe) is of type 2, where both absolute truth and the method of attaining that absolute truth exist, but neither can be specified with any finite quantity of information. Clearly the scientific method in a type 2 universe cannot be known in its entirety, as the positivists hoped. We can only know finite portions of it. Can we then specify an empirically testable way of developing legitimate parts of a nonfinite scientific method in a type 2 universe in which the complete method cannot ever be fully stated? I think this is possible. The philosophy of science in a type 2 universe should begin by being firmly grounded in the three basic concepts introduced earlier in this chapter: the collection, compression, and organization of information. These form the basis for a set of empirically testable criteria that can attain the *goal* set by the logical positivist school of philosophers, a set of criteria that can be used to distinguish valid scientific results from pseudoscience, even in a type 2 universe, where the positivists' *method* of attempting to completely specify scientific methods (i.e., completely specify the methods of collection and compression of information) fails because it cannot be accomplished with any finite quantity of information. In other words, even though the scientific method itself cannot be compressed into a finite quantity of information in a type 2 universe, we can develop empirically testable methods for deciding what is a valid component of the scientific method and what is not.

The first of these basic concepts, the collection of information, which is associated with major phase changes in science, is made empirically testable by the requirement that observations and experiments be repeatable. It might be argued (following Heraclitus) that no experiment or observation is exactly and perfectly repeatable, but perfection is an unreasonable standard. Absolute perfection is not attainable, so we need not be overly concerned with philosophers who demand it. Short of perfection, it is possible to set up reasonable standards for the repeatability of experiments and observations. These repeated experiments and observations should be carried out under reasonably varied and controlled conditions, preferably designed with a consciously skeptical attitude that involves reasonable attempts to falsify the original observation. Thus, Popper's vital insight into the importance of falsifiability of scientific results plays a central role in this portion of the philosophy of science in a type 2 universe.

The compression of information, as noted above, is associated with the development of theories and Kuhn's concept of paradigm shifts. And if the criteria for evaluating a theory can be grounded in the idea of compression of information then we can sidestep the main difficulty that faced the positivist school, the need to completely specify the methods of science. Instead, we need only ask a few empirically testable questions of a proposed theory:

1. Does it provide a greater compression of a given set of observed data than another theory?
2. Is it more accurate, closer to the observed data, and within the estimated uncertainties of that data?
3. Does it compress a larger range of data, generally including data that were not included in the original development of the theory and data that are not compressed as well by another theory?
4. Is there a range of parameter values in which a proposed novel compression of data produces significantly different values than another (and perhaps more generally accepted) compression scheme? Observations in this range of parameter values can be used to test the compression algorithms and falsify one or more of them.

Similar questions can be framed about the accuracy, reliability, and repeatability of observational and experimental information. Questions such as these can provide empirically testable methods of accepting or rejecting new data and theories in both a type 1 and a type 2 universe. Indeed, they provide a set of standards for empirically establishing and validating methodologies that are not limited by any assumption about the quantity of information needed for those methodologies. Although the complete methods that are used to compress information cannot be formulated with any finite quantity of information, once a particular compression is attained by any method at all ("anything goes") the compression can be easily and empirically tested. And the testing involved in answering the questions above is very similar to Popper's important concept that theories should be falsifiable, since comparison with observations is the essential feature of falsification. Popper's criterion of falsifiability of a theory is therefore a natural part of this procedure as well.

The further idea of organizing information that is incompressible and too vast to be assimilated is not something that is easily subject to an empirical test, but it remains an essential part of any branch of intellectual activity. All branches of science and mathematics would have to develop such skills in order to deal with large collections of incompressible data in a type 2 universe. It is unlikely that any scheme for data organization will last indefinitely. Such schemes will be refined and revised according to the tastes and interests of active researchers. The ultimate value of any organization scheme will be measured by the ease of finding information that a researcher does not carry in his own memory. This is an important parameter, but it is one that is not easy to quantify. Nevertheless, it is clear that the recent availability of enormous quantities of information on the internet and the World-Wide-Web, coupled with the existence of novel indexing and search engines, is having a profound effect on the way that information is stored and retrieved, especially in the sciences and mathematics. The recent trend toward on-line publishing in addition to or in place of classical journal publications marks a major change in the way that information is organized and accessed in the sciences and mathematics. The process of organizing information is being impacted by the computer revolution as profoundly as all the other components of science and mathematics.

Thus, the ideas of collecting, compressing, and organizing information can be used to form the core of a philosophy of science that has its essential elements subject to empirical testing and avoids many of the difficulties associated with the assumption that absolute truth and the methods for obtaining or approximating that absolute truth must be finite. However, although these methods are suitable for a type 2 universe, they work equally well in a type 1 universe because, as noted above, they make no assumption at all about the quantity of information needed for scientific methods. Indeed, even though the philosophy of science is radically different between these two cases, there would be essentially no difference in the *practice* of science in either case. We would still ask the same fundamental questions of a proposed theory. The only difference is that if we assume that the physical universe is a type 1, then we will expect the process to stop at some point when absolute truth is attained, but if we assume that the physical universe is a type 2, then we will not expect the process to come to an end. But the practice of science would be the same in either case, in sharp contrast to a type 3 universe, where no practice of science would be effective. In fact, using these concepts, the practice of science in a type 1 or a type 2 universe would not be significantly different from the practices and methods that are currently employed in the sciences.

These four questions represent processes or methods that are very close to what theorists today actually do. Indeed, they are close to I. I. Rabi's famous aphorism that "science means doing your damndest with your mind, no holds barred." Rabi's comment may apply even better in a type 2 universe than a type 1 universe, where the finite nature of the scientific method would represent a "bar" on some "holds." And even if the physical universe turns out to actually be a type 1 universe rather than a type 2 universe, there is no way to know that with any certainty using finite amounts of information, no way of knowing exactly when the process will stop (see the discussion in Robertson, 1998, chapter 2). But even though the question of whether the universe is a type 1 or 2 may be intrinsically interesting, in a practical sense it makes little or no difference to the everyday practice of science because the process of collection, compression, and organization of information would be essentially the same in either type of universe.

This system of empirically testing observation and compression schemes will not be perfect, of course. There are no guarantees of infallibility. There will probably be occasions where material that should be taken seriously will be rejected as pseudoscience, and more serious situations where pseudoscience is taken seriously for a time. But these aberrations can and should be eliminated over time as observations increase in quantity and accuracy, and a variety of novel compression algorithms are developed and tested by comparison against these observations.

Returning to Kitcher's question, "Can we legitimately view truth as a goal of science?," we can now see that the answer depends on the type of universe we are in. In a type 3 universe, the answer is no. But there is considerable evidence that we are not in a type 3 universe. In a type 1 or type 2 universe, the answer is yes. But then a related question immediately comes up: Does the process that is described here, of collecting, compressing, and organizing information, approach that ultimate truth and converge toward it? It seems to me that the answer to this question must be in the affirmative in a type 1 or a type 2 universe. Indeed, the absence of such convergence might be sufficient to define a type 3 universe. A universe in which such convergence does not happen would be effectively a type 3 universe, in that it could not be distinguished by human researchers from a genuine type 3 universe.

And the evidence that supports the hypothesis that we are not in a type 3 universe (i.e., the extraordinary success of conventional scientific methods) is vastly greater than any evidence that supports any proposed alternative to the hypothesis, particularly the alternatives proposed by the postmodernists. This working hypothesis also leads to good scientific practice and, so far, at least, it has led to an impressive string of successes in physics, mathematics, astronomy, and all across the sciences. And although we cannot know, outside of mathematics, whether the process will terminate or not, that is, whether the physical universe is a type 1 or a type 2 universe, we have seen that the answer to this question makes essentially no difference to the everyday practice of scientific methods, it only makes a difference in whether we expect the process to terminate or not, and this is not a question that we can expect to know the answer to in any finite amount of time. And since we cannot know the answer, we might as well maintain our working hypothesis that the universe is a type 2 universe indefinitely, since, as noted above, there is some evidence for this hypothesis as well.

Stephen Wolfram recently suggested that there is a "single simple algorithm that, in effect, generates all the rules of physics and everything else" (Levy, 2002, p. 148). Wolfram goes on to suggest that this simple rule can be expressed in a few lines of computer code, perhaps three or four lines, expressed in his symbolic-algebra computer program that he calls Mathematica. He has outlined these ideas in some detail in his book *A New Kind of Science* (Wolfram, 2002). If Wolfram's arguments hold up, they would represent an incredible level of compression of information, perhaps the ultimate compression, a "theory of everything" for the entire universe compressed into a few lines of computer code. It seems to me quite likely that Wolfram is describing a powerful method of compression of information. But whether it compresses the entire universe, or merely some very interesting and perhaps infinite subset of the universe, seems to me to be a critical question that cannot be readily answered. The principal difference between Wolfram's argument and the one presented here is that, while we both believe that the universe can be expressed in lines of Mathematica code, I think that there is a good chance that the number of lines of code needed is not at all small, indeed, that it is not finite. Elementary concepts of conservation of information under logical operations, coupled with the known infinity of compressed information that is needed for mathematics, seem to make Wolfram's hypothesis unlikely, if not untenable.

Wolfram is not alone in thinking that the physical universe is a type 1 universe, whose TOE can be expressed with a small amount of information. Steven Weinberg also asserts that physicists are approaching a final theory, consisting of "a few simple general laws" (2001, p. 13). E. O. Wilson also expects the sciences and the humanities to evolve toward a "Magellanic voyage" that will encompass all knowledge (1998, p. 268). But if the physical universe is a type 2 universe, then these ideas will all run afoul of Chaitin's concept of conservation of information, very much as attempts to produce perpetual motion machines in the nineteenth century ran afoul of the law of conservation of energy. There is no way that an infinity of incompressible information can be expressed with any finite number of "simple general laws" (axioms).

Alfred North Whitehead's famous dictum "Seek simplicity and distrust it" is relevant here. Whitehead's pronouncement is an essential update to Occam's razor appropriate for a type 2 universe: "Seek simplicity" is simply a restatement of Occam's razor, but the second half, "distrust it," is a necessary addendum for a type 2 universe, in which no finite statement of scientific truth is completely adequate.

Thus, we now know, at least in mathematics and in the physical universe as well if it is a type 2 universe, that in place of Wilson's Magellanic voyage that will encompass all knowledge, we will instead have an unending process of collecting, compressing, and organizing information. The long search by mathematicians, physicists, biologists, philosophers, theologians, and others for absolute certainty and final, absolute knowledge of the mathematical universe (and probably the physical universe as well) is fundamentally hopeless. Yet in exchange for this certainty and finality we actually have something much better: an unending adventure. The sciences and the humanities are seen to involve unterminating research. And if current experience is any guide, that research will often take surprising turns that go off in new and unexpected directions, often directions of spectacular wonder and beauty and power.

To borrow Darwin's famous closing words, "There is a grandeur in this view of life," the view that life is an unending process of exploration, an unterminating sequence of discovery and adventure, but it is a grandeur that should be approached with a degree of humility. We must leave behind the hubris that would seek absolute certainty and final, complete knowledge, and instead, be content with the unending adventure of study and learning—the collection, compression, and organization of information.

9

CONCLUSION

Phase Changes and Information Revolutions

The main theme of this book can be summarized simply: The history of science and mathematics has been marked by phase changes driven by technological innovations that gave us the ability to see things that had never been seen before. The telescope in astronomy and the microscope in biology are perhaps the most familiar examples of such critical technological innovations. Today's computer technology is expanding our ability to see things just as the telescope and the microscope did in the past, but this new computerized power to see things is taking place on a much larger scale and over a far wider range of fields of study than has ever happened before. The increases in resolution and sensitivity that are being provided by this new computerized instrumentation are often many orders of magnitude larger than the comparable increases for instruments invented in the past. These new computerized instrumental capabilities are driving the largest and most important phase changes ever experienced in every field of science. Simultaneous revolutions of this breadth and magnitude are completely unprecedented in the entire history of science and mathematics.

At the same time the computer is also the most important tool ever invented for helping us to understand the new discoveries that are being made. Computers give us extraordinary new power to develop novel theories, new compressing algorithms that reduce massive amounts of raw data to comprehensible patterns. With computer technology we will be able to develop mathematics much further and faster than we could ever have done without computers. Whereas theorists in the precomputer era were restricted largely to linear problems and very simple nonlinear problems with special symmetries, computers give us extraordinary new power to handle the really difficult theoretical problems, the general nonlinear problems that are found throughout mathematics, physics, astronomy, biology, and through the rest of the sciences.

Finally, the information theory that provides the theoretical basis for much of the computer revolution also gives us important new perspectives and novel insights into the basic philosophy of science. The concepts of collection, compression, and organization of information allow us to outline a philosophy of science that avoids the difficulties that bedeviled such philosophies in the past, where the vital distinction between a type 1, 2, and 3 universe was not recognized, and indeed, the information-theoretic concepts that are needed to define these possible universes had not been developed.

Only a small fraction of this vast story has been recounted here. Indeed, if it were possible to contain the entire story within a single book, then it would be a lot less interesting than it actually is. Additional examples of phase changes abound in other fields of research and are not at all difficult to find. In the course of writing this book, I found it was nearly impossible to read through any recent issue of a major scientific journal such as *Science* or *Nature* without finding several quotes or examples of things that could be used somewhere in this book. The sheer volume of such examples was so huge that most of them simply could not be used here. Medical technology, for example, experienced a phase change in the early twentieth century with the development of X-ray techniques that allowed surgeons to see inside a living body without cutting into it. Today computer technology has been essential to the development of a variety of new noninvasive ways of seeing inside living bodies, including computerized axial tomography (CAT) scans, positron emission tomography (PET) imaging, nuclear magnetic resonance (NMR) imaging, and ultrasound techniques, for example.

Engineering is another field in which computers have caused phase changes. Algorithms such as finite-element methods allow engineers to do the massive calculations that are needed to see and understand the forces, stresses, and strains involved in the structure of a bridge or an aircraft. In the precomputer era, these calculations were done as completely as possible with pencil-and-paper calculations, and then large and inefficient safety margins had to be built in to compensate for the inadequacy of the calculations. Today these calculations can be done far more accurately and completely to produce designs for buildings and aircraft and other products that are much more efficient than conventionally designed products and at the same time have more accurately calculated safety margins. And, of course, in electrical engineering the design of new computer processors and other microchips would be impossible without computer aid. Modern computer chips that have tens of millions of transistors and comparable numbers of other electronic components have attained such a level of complexity that no one would even think of attempting to design them by hand anymore.

This list could be continued indefinitely. The ability of computers to enhance our ability to see and understand things extends not only across all areas of science, mathematics, and engineering, but it also extends into the humanities. For example, Stanford University operates a Center for Computer Assisted Research in the Humanities, as well as a Center for Computer Research in Music and Acoustics. These centers are developing computerized databases of music as well as collections of software tools designed to analyze the contents of these musical databases. At the same time, computers are enabling historians to search, sort through, and organize historical documents in unprecedented quantity. And in literature, for another example, computers are providing new tools for analyzing a writer's works by allowing researchers to easily produce word counts and word usage statistics. These statistics may reveal new information about that writer's ideas and thought patterns and their development across time. And a recent book about different systems of writing faced the problem of having to mix text in conventional English with Cuneiform, Hieroglyphic, Chinese, Mayan, Runic, Georgian, Brahmi, Algonquin, shorthand, and dozens of other scripts, including those that were developed to specify music and dance. Daniels and Bright (1996, p. xxxvi) comment that

> a book like this would have been impossible ten or even five years ago. To typeset all the characters of all the languages involved would have required the labor of numerous specially trained compositors at a number of different printing houses. But the implementation of multiscript technologies by Apple Computer has made it feasible, if not exactly simple, to prepare the entire manuscript electronically.

This extraordinary breadth of the impact of computer techniques is perhaps more astonishing than the magnitude and power of those techniques. Computer technology can create revolutions essentially everywhere that information is handled in any form. We should expect to see revolutions in every field that made use of printed books and other written information in the precomputer era. It is important to understand this broader context of the computer revolution, and I would like to briefly address some of its wider philosophical implications. To place the computer revolution in science and mathematics into its proper perspective, we need to see it as a very important component of today's information revolution but still as only a part of this new revolution.

There is a deep and fundamental reason why information revolutions such as the computer revolution are so very important. Processing information is central to life itself; all life forms must process information in some fashion. Even bacteria collect, process, and respond to information about the chemical and physical state of their immediate environment. And multicellular forms of animal life (metazoans) have special needs and problems that center on information processing. There would be no point (no fitness advantage) in developing sensory apparatus (e.g., eyes) that is separate from muscle tissues without having some means of passing information from the sense organs to the muscles. Metazoans must first collect and interpret specific information from their sensory apparatus, usually information such as: (1) Over here is something I want to eat or drink; (2) over there is something that wants to eat me; (3) somewhere else is a good-looking potential mate; and (4) it's getting a little chilly here. Having collected and evaluated this information, the organism must then respond effectively by passing this information along to muscle tissues in a manner that induces an appropriate form of muscle activity (pounce, flee, wink, and so forth). Metazoans have developed nervous systems and brains that perform this information-processing function.

With one exception, metazoans use their brains and nervous systems to deal almost exclusively with questions of this nature. Our own species, *Homo sapiens,* is the only one that ever goes substantially beyond processing information at this level (at least some individuals do, some of the time). One of the principal defining characteristics, perhaps *the* principal defining characteristic of the human race from its very inception has been its extraordinary and unprecedented ability to handle information. We don't run faster than other organisms, and we don't have sharper teeth or thicker hide, but we do collect, process, and exchange information more effectively and in larger quantities than any other organism that has ever lived.

The human race has always been distinguished primarily by its unique capabilities for producing, storing, and distributing information, originally through the use of spoken language, and then with the use of written and printed language. As I outlined in my previous book (Robertson, 1998, chapter 1), these three information technologies, language, writing, and printing, each caused an information explosion in the past, and each of these information explosions was associated with a major revolution in human society. The invention of language marked the origin of the human race itself. The invention of writing marked the beginning of classical civilization. And the invention of printing marked the beginning of modern civilization.

We might describe each of these three revolutions in terms of a phase change, but in doing that we should recognize that a phase change in civilization is not exactly the same thing as a phase change in science and mathematics. Both types of phase change share the property that extrapolation of the previous behavior of the system does not give a clue to the present behavior of the system. But whereas a phase change in science and mathematics was associated with a novel ability to *see* things, a phase change in civilization is more closely associated with a novel ability to *do* things, as with the ability to print newspapers and electoral ballots, for example.

With this definition of a phase change in civilization we are now in a position to assert that information explosions were the central driving force behind the most important phase changes in the history of humanity. The simple underlying reason for this is that civilization is fundamentally information-limited, limited by its ability to collect, compress, and organize information. Civilization is information-limited for the same reason that it is energy-limited: Both energy and information are governed by a conservation law; information must be extracted from our immediate environment much as coal and oil are extracted. The invention of the computer marks not only the next step in this sequence of phase changes in information-limited civilization but also by far the largest step in the sequence, the largest information explosion in history.

Ideally, this assertion that civilization is information-limited and that information revolutions are the principal cause of phase changes in civilization should be subjected to the most stringent tests before being accepted. For the first two of the three information revolutions in history, the ones that followed the inventions of language and writing, these tests are very difficult because these two revolutions are too far removed in time to allow close study. Very little detailed information survives today about the invention of language and writing. But the information explosion that followed the invention of the printing press is significantly different in this regard; we do have enough information about it to be able to study it in some detail. And the most important question that we can ask about this information revolution is quite simple: How can we devise an experimental test of the idea that the printing press (and its corresponding information explosion) was the critical element that caused the phase change in civilization that we now call the Renaissance, the beginning of modern civilization, so that without this information explosion our modern civilization would not have developed? The results from such a test could give us important new perspectives on the impact of the newest information revolution, the one caused by computer technology.

In physics or biology we might try to perform such a test by repeating the experiment under different conditions. In other words, we could try to set up a civilization that was identical to that of Europe in 1440 and then suppress or eliminate the invention of the printing press. We would then observe the differences in the subsequent development of European civilization with and without the printing press. Obviously, it is not possible to carry out this type of designed sociological experiment. But we actually have the next best thing. The next best thing would be an example of two (or more) civilizations that are very similar, with one perhaps a little ahead of the other. If one of them adopts the printing press (preferably the one that is a little backward) and the other one does not, we could then observe the differences in the subsequent development of the two civilizations. Something very close to this proposed experiment was actually carried out by a pair of civilizations half a millennium ago. In the early fifteenth century, the civilizations in Europe and in the Middle East were very similar in many of their fundamental characteristics, perhaps as similar as it is possible for two different civilizations to be. But they took radically different courses in regard to the printing press. And today we can investigate the consequences that resulted from those fateful choices by examining the major differences between these two civilizations today.

Much of the early similarity between these two civilizations stemmed from the fact that they shared many of the same roots in the early civilizations of the eastern Mediterranean region. And by the early fifteenth century civilization in the Middle East was substantially ahead of the one in Europe in many important areas of mathematics, science, literature, art, architecture, and other important features of a civilization. This early dominance of the Middle Eastern civilizations is reflected today in the vocabulary of the English language, which contains many words that were derived from the Arabic language. These include particularly words and names that are used in mathematics, science, and architecture, words such as algebra, algorithm, alkali, alcohol, alembic, alcove, Aldebaran, and Algol. Even the decimal numbering system that is essentially in universal use today was called Arabic after its introduction to Europe by Leonardo of Pisa (aka Fibonacci) because he had learned it from Middle Eastern scholars. Decimal numbering was probably invented in India, but the mathematicians of the Middle East were using it long before European mathematicians were aware of it.

Princeton's Bernard Lewis recently attempted to formulate a conventional explanation for the differences found between Middle Eastern and European civilizations today (Lewis, 2002). He cited an impressive list of problems with Middle Eastern civilizations, including such things as repressive governments, suppression of individual freedom and women's rights, and religious intolerance. Lewis constructed an impressive argument with only one major difficulty: It is easy to make the case that all the things that he cites as problems in the Middle East were also serious problems in European civilization in the fifteenth century and for a considerable length of time afterward.

Lewis goes on to suggest that the cause of the difference is to be found in Europe rather than the Middle East, but he does not even speculate on the nature of the factor that changed Europe (Lewis, 2002, pp. 155–156):

> An argument sometimes adduced is that the cause of the changed relationship between East and West is not a Middle-Eastern decline but a Western upsurge—the Discoveries, the scientific movement, the technological, industrial and political revolutions that transformed the West and vastly increased its wealth and power. But these comparisons do not answer the questions; they merely restate it—Why did the discoverers of America sail from Spain and not a Muslim Atlantic port, where such voyages were indeed attempted in earlier times? Why did the great scientific breakthrough occur in Europe and not, as one might reasonably have expected, in the richer, more advanced, and in most respects more enlightened realm of Islam?

Lewis fails to identify the one very important way in which these two civilizations were demonstrably different, almost diametrically opposite: The civilization in Europe adopted the printing press in the fifteenth century, while the civilization in the Middle East actively suppressed the use of printing. Although Lewis is clearly aware of the absence of printing in the Middle East (Lewis, 2002, pp. 142–143), he is apparently not aware of its critical importance to the major changes that took place in Europe during this time period. Printing had a direct impact both on the fact that Columbus sailed from Spain and on the great scientific and technological and political breakthroughs that occurred in Europe.

The printing press was introduced by Gutenberg in the 1440s and exploded across Western Europe in the second half of the fifteenth century, very similar to the way that computer technology exploded in the second half of the twentieth century, albeit on a smaller scale. By the year 1500 printing presses were established in major cities across Europe. William Caxton in London and Aldus Manutius in Venice were among the more famous successors to Gutenberg. Oxford University Press (the publisher of this book) was founded as early as 1478. (Many of the direct effects of those newly printed works in Europe were surveyed in Robertson [1998, chapter 1].) The explosion of information that followed the invention of the printing press was of central importance to the revolutions that are collectively termed the "Renaissance," the phase change that is rightly regarded as the birth of modern civilization. For example, by the 1480s Columbus was in possession of printed copies of Pliny and Marco Polo along with other works on geography that were essential not only for the development of his ideas about reaching Cathay by sailing across the Atlantic but also, equally important, were essential to his ability to obtain critical support from a skeptical court in Spain. Again, in the early 1500s Martin Luther sold hundreds of thousands of printed copies of his writings. Dickens (1966, p. 51) argues that access to the printing press was the principal difference between Luther and his predecessors in the Wycliffite and Waldensian movements. And Harvard's Owen Gingerich has this to say about the dependence of Copernicus (and others) on the technology of the printing press (1975, p. 202):

> The early 1500s were times of vast changes. Oceanic navigators opened new continents and an age of exploration. Da Vinci and Durer coupled mathematics with art to capture new harmonies of proportion and perspective. Martin Luther successfully set into motion a reformation of the church. . . . Meanwhile, the explosive spread of printing

> with movable type beginning in the 1450s fanned the sparks of all these movements, including the reform of astronomy. Without printing, Copernicus would have been deprived of the vast majority of his source materials. Even five decades earlier, he could not easily have found the requisite information that built his *De Revolutionibus* into the greatest astronomical treatise of its century. And without printing, his manuscript might have languished, virtually forgotten, on the shelves of the cathedral library.

The other item cited by Lewis, the "political revolutions that transformed the West," was also critically dependent on the technology of the printing press. The development of modern democratic systems of government in the West could not have taken place without printing. Not only do electoral ballots need to be printed, but also the entire continental-scale democratic political process was critically dependent on the ability to disseminate information to a broad electorate through printed newspapers and other publications.

In sharp contrast both the Ottomans and the Mamelukes, the principal Middle Eastern powers of the period, suppressed the technology of the printing press for centuries. In 1485 the Ottoman Sultan Bayezid II outlawed not only the printing press but even the simple possession of printed matter, and his son and successor Selim I renewed the ban. As Lindner noted (1998, p. 94):

> The reluctance of Muslim powers to allow printing from movable type is well-known. The first books officially printed in such fashion appeared from an Istanbul press in 1729, and the manager of the shop was a Hungarian convert to Islam and Ottoman official called Ibrahim Muteferrika. The output of his press was limited, seventeen titles before his death in 1745, each produced in a tirage ranging from twelve hundred down to five hundred copies, with most titles at the lower figure. The circulation figures suggest, then, that these books did not range widely and deeply into Ottoman society, even literate Ottoman society.

Seventeen titles is a painfully small effort almost three hundred years after Gutenberg and hardly constitutes an information explosion.[1] And to find the first printed works in Turkish, you have to turn to Europe (Lindner, 1998, p 95):

> The printing of works in Islamic languages and the Arabic alphabet occurred in Europe substantially before it began in the Near East. For example, the first work printed in Turkish using movable type appeared in Paris in 1615.

The first modern print technology to catch on in the Middle East was lithography, which was ideally suited to printing the Arabic alphabet (Lindner, 1998, p. 98):

> Discovered by G. A. Senefelder in 1796, the lithographic process was first used in Turkey in 1801. For many of the scripts as well as certain kinds of texts (especially the Koran), lithography remained the process of choice until recently. Certain scripts attached themselves through custom to certain genres of literature, and these variants meant that a print shop might have either to limit the breadth of its production or to make a substantial investment in punches and matrices at the very beginning. Neither of these options was attractive, especially since the problem of curving lines and intersection letters remained unsolved.
>
> The possibilities of Arabic script, then, allowed for pages that simply could not be reproduced with movable type.

Of course lithography is essentially a refinement of block printing, and although it could handle the problems of Arabic script without difficultly, it does not have the flexibility of printing with movable type. Lindner describes other reasons for the aversion to printing by the ruling powers in the Middle East, reasons that ranged from theological objections to the mass production of holy writ to Luddite-like objections from professional copyists and calligraphers. There was even a major difficulty with the production of paper for printing—the entire Middle East was short on trees for paper production. Paper had to be imported from Europe.

There is no more distinct difference between European and Middle Eastern civilizations in the period from the fifteenth century to the nineteenth century than the inundation of Europe with printed works and the corresponding vacuum of printed material in the Middle East. In the information vacuum that resulted from the absence of printing with movable type in the Middle East, there was no phase change in civilization there comparable to the Renaissance in Europe. Civilization in the Middle East remained largely unchanged from the fifteenth through the nineteenth centuries.

This difference between the development of European and Middle Eastern civilizations in the period from the fifteenth century through the nineteenth century provides us with a fascinating and probably unique sociological experiment that is dominated by a single factor, the presence or absence of the printing press and the associated information explosion that was generated by the printing press in Europe.

And the difference in the subsequent development of the two civilizations—a renaissance in the civilization that exploited the printing press and stagnation in the civilization that did not—gives us critical evidence that bears directly on the importance of the next information revolution, the one being touched off by computer technology today, which is one of the central themes of this book. To understand the magnitude of the impact of the computer revolution, we need only compare European civilization today with present-day Middle Eastern civilization (or with Europe in the fifteenth century) and then multiply the difference by about a half-dozen orders of magnitude to account for the difference in the size of the two information explosions.

Similar patterns must have existed for the two earlier information explosions, the phase changes that followed the invention of language and the invention of writing, but we know almost nothing of the details of those early information explosions. For the first phase change, that occurred some fraction of a million years ago, one line of primates in Africa invented language and developed into what we now call the human race. Then about 5,000 years ago, different groups in Mesopotamia, Egypt, the Indus Valley, and China invented writing and produced a second phase change, one that resulted in the creation of what we refer to as the first or classical civilization.

With the invention of writing, one of these early civilizations came startlingly close to having enough information in hand to produce a scientific and industrial revolution. The greatest concentration of information in antiquity was found in the great library of Alexandria in Egypt, and the scientists and mathematicians associated with the great library developed many of the fundamental elements of a scientific revolution. Euclid laid the foundations for all of mathematics in a form that remains recognizable after two millennia. Archimedes understood many of the fundamental concepts of Newton's calculus and used them to calculate the area of a circle, the area under a parabola, the volume of a sphere, and the area of the surface of a sphere. Ptolemy developed models for planetary motions whose accuracy was not surpassed for more than a dozen centuries. And Heron was even experimenting with steam engines.

But the great library was too small and vulnerable to destruction. And access to it was very limited. Geniuses such as Archimedes who were not in residence in Egypt had to make difficult and dangerous journeys just to visit the library. The real scientific and industrial revolution had to wait for the invention of the printing press, when the production of information first attained production rates that exceeded the capacity of any single library. The effects of this information explosion were summarized by Gingerich, earlier, and outlined in more detail in Robertson (1998, chapter 1).

As noted above, the most important defining characteristic of the entire human race, the thing that makes us significantly different from all the other organisms on the planet, is our unique ability to handle information. This novel ability to handle information has defined the human race since the invention of language, and it received a significant boost with the inventions of writing and printing. Today computer technology is producing the largest information explosion of all time, quantitatively much larger than the previous three information explosions combined (see the discussion in Robertson, 1998, pp. 20–24). The invention of the computer is therefore one of the pivotal events in the history of the human race, perhaps the defining moment in that entire history. The new information revolution that is being generated by computer technology is already creating an even larger phase change in civilization than the information explosion that was touched off by the printing press. And the revolutions of the computer era will be centered in the sciences and mathematics, as outlined in the earlier chapters of this book, just as the printing press touched off a scientific revolution in the sixteenth through the eighteenth centuries. This book has explored only a small part of that much larger story.

The information-theoretic viewpoint that has been developed here, that was made possible by the development of information theory in the mid- to late-twentieth century, has implications far beyond merely revolutionizing the philosophy of science. It has a similar effect on almost every part of philosophy, and it substantially changes the way we think about such things as the progress of civilization. Indeed, nothing looks quite the same after it has been analyzed from an information-theoretic viewpoint. The full implications of the applications of information theory to classic problems in philosophy are only beginning to be explored. The idea that civilization is information-limited provides an important new perspective on the development of civilization. All areas of civilization, from mathematics through all of the sciences and engineering, to the humanities and the arts, are concerned with collection and organization of information, and the sciences add the important concept of compression of information. But we have seen that nearly all information is not compressible, and that therefore many areas of study must confine themselves to collection and organization. (Indeed, the collection and organization of information is sometimes denigrated, not very intelligently, with terms such as "stamp collecting." An information-theoretic viewpoint makes clear why such denigration is inappropriate, and why the collection and organization of information is a vital component of research.)

This novel information-theoretic perspective suggests a possible new way to define the sciences. If we consider all branches of intellectual endeavor, including the hard sciences, the soft sciences, the humanities and the fine arts, to be varieties or forms of art, then the sciences could be defined as the art forms that are concerned with the rare but important classes of information that are compressible. This definition would be very close to historical uses of the term "science" and would allow an empirical test of whether an area of study should be called "science." Further study would be needed to determine exactly how useful such a definition might be. (It would probably be necessary to exclude mathematically trivial forms of compressibility such as are used to create "zipped" files or .jpeg graphics files in computer storage. Not all compression algorithms are of equal value in providing scientific insight.)

These philosophical developments that relate the philosophy of science and mathematics to the modern theory of information lead us to an obvious question: How well does this new philosophical perspective enable us to respond to Hilbert's challenge in the first chapter, "to lift the veil behind which the future lies hidden?" Admittedly, we have not "lifted the veil" as far as we might desire. But to continue the analogy from chapter 1 of exploring unknown waterways by canoe, we have determined that not only are there waterfalls ahead, but we are today in the very midst of some of the largest waterfalls ever seen, in every direction, and we should prepare for a wild ride.

The revolutions in science and mathematics that are taking place today are some of the most exciting adventures we have ever experienced. The voyages of the sixteenth century, the "age of discovery," and the scientific revolutions of the sixteenth through the eighteenth centuries or even the twentieth century pale by comparison to the excitement of the flood of discoveries being made today. And, just as in that earlier age of discovery, the new explorations and discoveries that are being made today will shape the course of civilization through the future. Researchers today are more fortunate than any others in all history. They are living in a period where the capabilities of instrumentation are exploding, when data can be collected at rates that earlier generations could only dream of, and the computer hardware and novel analytic and theoretical methods needed to understand the nonlinear patterns in that data are available for the first time. Wordsworth captured the soul of the present epoch in the history of research with his verse that was cited at the beginning of the book and which we echo here: “Bliss was it in that dawn to be alive.”

NOTES

Chapter 1

1. Technically, the freezing process is a bit more complicated than this. For example, water can enter a metastable or supercooled state before it suddenly crystallizes. But these and other technical details are not important for the discussion here.

Chapter 2

1. Readers who are interested in learning more about the SETI at home project or participating in it are invited to look at the Web site: http://setiathome.ssl.berkeley.edu/.

Chapter 6

1. Eratosthenes' famous measurement depended critically on the assumption that the Earth has a spherical shape, as well as on the assumption that the distance to the Sun is much larger than the radius of the Earth. If the Earth were flat, then Eratosthenes' shadows could be explained by assuming that the Sun is at a distance about 6,000 kilometers above the flat Earth. The difference between this flat-Earth model and a spherical model could have been checked by making measurements of shadow lengths at more than two different latitudes, but to my knowledge no one in antiquity ever did that, partly because after Aristotle's time no one seriously questioned the spherical shape of the Earth.

Chapter 8

1. Experts in the area will recognize that I have neglected several important concerns here such as bandwidth limitations, error levels, and roundoff errors. These and other related issues do not directly affect the arguments made here.

2. There is a difficulty with terminology here. For a given bitstring, the quantity that is of interest is the length of the shortest computer program that will generate that bitstring. I have called that quantity the "information content" of the bitstring, but of course the word "information" comes with a lot of baggage that does not apply here. Chaitin uses the word "complexity" to mean the same quantity, but that word also comes with a fair amount of baggage; its conventional usage does not generally suggest a numerical quantity. Perhaps it would be best to define a new word: The "floop" of a bitstring is the length of the smallest program that would generate the bitstring. If the reader is comfortable with this then he or she is more than welcome to substitute the word "floop" for the terms "information content" or "quantity of information." (Apologies to D. Hofstadter for borrowing the word "floop," which also comes with some baggage [1979, p. 406].)

3. Technically, of course, it is a combination of mass and energy that is conserved. We will consider mass to be a form of energy and simply speak of conservation of energy.

4. It is not easy to specify the exact minimum number of axioms needed for Goedel's theorem to apply. Roughly, if one has enough axioms to define the integers and the arithmetical operations of addition, subtraction, multiplication, and division, then this is sufficient.

5. The word "dominance" here means simply that today you almost never find invertebrates above vertebrates in the food chain. Fish- and bird-eating insects and spiders and fish-eating molluscs are among the rare exceptions. There are even fish-eating Cnidarians. Interestingly, nearly all these examples involve the use of toxins or poisons, that is, chemical rather than physical dominance.

6. It is possible to make a further breakdown into, for example, a type 1.5 universe, where the ultimate truth is finite but the scientific methods are not, but I do not see much need to break things down that far.

7. Michelson may have been quoting Lord Kelvin. See the discussion in Weinberg, 1992, p. 13.

Conclusion

1. Actually, a printing press had been set up in Istanbul in 1493 by Jewish refugees from Spain. The ban on printing did not extend to works in Hebrew and other non-Arabic or Turkish languages printed by non-Muslims.

REFERENCES

Agnew, D.C. 2002. History of Seismology. In *International Handbook of Earthquake and Engineering Seismology,* Part A, ed. W. H. K. Lee, H. Kanamori, P. C. Jennings, and C. Kisslinger, 3–11. New York: Academic Press.

Andrade, E. N. da C. 1964. *Rutherford and the Nature of the Atom.* Garden City, N.Y.: Doubleday.

Arkani-Hamed, N., S. Dimopoulos, and G. Dvali. 2002. Large Extra Dimensions: A New Arena for Particle Physics. *Physics Today* 55, no. 2: 35–41.

Bailey, D. H., and J. M. Borwein, 2001. Experimental Mathematics: Recent Developments and Future Outlook. In *Mathematics Unlimited—2001 and Beyond,* ed. B. Enquist and W. Schmid, 51–66. Berlin: Springer.

Baker, W. E., et al. 1995. Lidar-Measured Winds from Space: A Key Component for Weather and Climate Prediction. *Bulletin of the American Meteorological Society* 76, no. 6: 869–886.

Balsley, B. B., M. J. Jensen, and R. G. Frehlich. 1998. The Use of State-of-the-Art Kites for Profiling the Lower Atmosphere. In *Boundary Layer Meteorology,* ed. J. A. Garratt and P. A. Taylor. Dordrecht: Kluwer Academic.

Barrow, J. D. 1998. *Impossibility: The Limits of Science and the Science of Limits.* Oxford: Oxford University Press.

Bell, G. I. 1988. Preface to Computers and DNA. In *Proceedings Volume VII, Santa Fe Institute Studies in the Sciences of Complexity,* ed. G. I. Bell and T. G. Marr, xiii–xiv. Redwood City, Calif.: Addison Wesley.

Bourguignon, J-P. 2001. A Basis for a New Relationship Between Mathematics and Society. In *Mathematics Unlimited—2001 and Beyond,* ed. B. Enquist and W. Schmid, 171–188. Berlin: Springer.

Bronstein, M. 1997. *Symbolic Integration I, Transcendental Functions, Algorithms and Computation in Mathematics, Vol. 1.* Berlin: Springer.

Butler, D. 2001. Are You Ready for the Revolution? *Nature* 409: 768.

Chaitin, G. J. 1970. On the Difficulty of Computations. *IEEE Transactions on Information Theory* IT-16: 5–9.

Chaitin, G. J. 1975. Randomness and Mathematical Proof. *Scientific American* 232, no. 5: 47–52.

Chaitin, G. J. 1990. *Information, Randomness & Incompleteness, Papers on Algorithmic Information Theory.* Singapore: World Scientific.

Chaitin, G. J. 1999. *The Unknowable.* Singapore: Springer.

Chaitin, G. J. 2002. *Conversations with a Mathematician.* London: Springer.

Chalmers, A. F. 1990. *Science and Its Fabrication.* Minneapolis: University of Minnesota Press.

Cohen, A. M. 2001. Communicating Mathematics Across the Web. In *Mathematics Unlimited—2001 and Beyond,* ed. B. Enquist and W. Schmid, 283–300. Berlin: Springer.

Cohen, J. 2001. The Proteomics Payoff. *Technology Review* 104, no. 8: 54–60.

Daniels, P. T., and W. Bright. 1996. *The World's Writing Systems.* Oxford: Oxford University Press.

Davis, P. J. 1998. Fidelity in Mathematical Discourse. In *New Directions in the Philosophy of Mathematics,* ed. T. Tymoczko, 163–176. Princeton, N.J.: Princeton University Press.

Dickens, A. G. 1966. *Reformation and Society in Sixteenth-Century Europe.* New York: Harcourt, Brace and World.

Dunham, W. 1990. *Journey through Genius: The Great Theorems of Mathematics.* New York: Wiley.

Dyson, F. 1985. *Infinite in All Directions.* New York: Harper and Row.

Eisenstein, E. L. 1979. *The Printing Press as an Agent of Change.* 2 vols. Cambridge: Cambridge University Press.

Feder, T. 2002. Astronomers Envision Linking World Data Archives. *Physics Today* 55, no. 2: 20–22.

Feyerabend, P. K. 1975. *Against Method.* Norfolk: Thetford Press.

Geymonat, L. 1965. *Galileo Galilei: A Biography and Inquiry into His Philosophy of Science,* S. Drake, trans. New York: McGraw-Hill.

Gingerich, O. 1975. Copernicus and the Impact of Printing. In *Vistas in Astronomy,* 17, ed. A. Beer and K. A. Strand, 201–207. Oxford: Oxford University Press.

Grafton, A. 1997. *The Footnote: A Curious History.* Cambridge, Mass.: Harvard University Press.

Gray, J. J. 2000. *The Hilbert Challenge.* Oxford: Oxford University Press.

Greene, B. 1998. *The Elegant Universe.* New York: W. W. Norton.

Gross, P. R., and N. Levitt. 1994. *Higher Superstition: The Academic Left and Its Quarrels with Science.* Baltimore, Md.: Johns Hopkins University Press.

Gunton, J. D., M. San Miguel, and P. S. Sahni. 1973. The Dynamics of First Order Phase Transitions. In *Phase Transitions and Critical Phenomena,* vol. 8, ed. C. Domb and J. L. Lebowitz, 269–466. New York: Academic Press.

Hallam, A. 1992. *Great Geological Controversies.* Oxford: Oxford University Press.

Hamilton, E. 1965. *Three Greek Plays.* New York: W. W. Norton.

Hanfling, O. 1981. *Logical Positivism.* Oxford: Basil Blackwell.

Hayes, B. 2001. The Weatherman. *American Scientist* 89, no. 1: 10–14.

Hitchen, N. 2001. Global Differential Geometry. In *Mathematics Unlimited—2001 and Beyond,* ed. B. Enquist and W. Schmid, 577–592. Berlin: Springer.

Hofstadter, D. 1979. *Gödel, Escher, Bach, an Eternal Golden Braid.* New York: Random House.

International Human Genome Sequencing Consortium. 2001. Initial Sequencing and Analysis of the Human Genome. *Nature* 409: 860–921.

Irion, R. 2002. Tuning in the Radio Sky. *Science* 296: 830–831.

Jayawardhana, R. 2002. The Age of Behemoths. *Sky and Telescope* 103, no. 2: 31–37.

Katinka, M. D., et al. 2001. Genome Sequence and Gene Compaction of the Eukaryote Parasite *Encephalitozoon cunculi. Nature* 414: 450–453.

Kauffman, S. A. 1993. *The Origins of Order: Self-Organization and Selection in Evolution.* New York: Oxford University Press.

Kitcher, P. 1993. *The Advancement of Science: Science without Legend, Objectivity without Illusions.* New York: Oxford University Press.

Knowles Middleton, W. E. 1969. *Invention of the Meteorological Instruments.* Baltimore, Md.: Johns Hopkins University Press.

Kuhn, T. 1970. *The Structure of Scientific Revolutions.* 2d ed. Chicago: University of Chicago Press.

Lakatos, I. 1978. History of Science and its Rational Reconstructions. In *Imre Lakatos, Philosophical Papers, Volume 1: Methodology of Scientific Research Programmes,* ed. J. Worrall and G. Currie, 102–138. Cambridge: Cambridge University Press.

Langtangen, H-P., and A. Tveito. 2001. How Should We Prepare the Students of Science and Technology for a Life in the Computer Age? In *Mathematics Unlimited—2001 and Beyond,* ed. B. Enquist and W. Schmid, 809–826, Berlin: Springer.

Lax, P. D. 1989. The Flowering of Applied Mathematics in America. *SIAM Review* 31: 533–541.

Lederman, L. M., with Dick Teresi. 1993. *The God Particle: If the Universe Is the Answer, What Is the Question?* Boston: Houghton Mifflin.

Lee, W. H. K., H. Kanamori, P. C. Jennings, and C. Kisslinger, eds. 2002. *International Handbook of Earthquake and Engineering Seismology, Part A.* New York: Academic Press.

Levy, S. 2002. The Man Who Cracked the Code to Everything. *Wired* (June): 132–147.

Lewis, B. 2002. *What Went Wrong? Western Impact and the Middle Eastern Response.* Oxford: Oxford University Press.

Lindner, R. P. 1998. Icons among Iconoclasts in the Renaissance. In *The Iconic Page in Manuscript, Print, and Digital Culture,* ed. G. Bornstein and T. Tinkle, 89–108. Ann Arbor: University of Michigan Press.

Mackenzie, D. 1995. The Automation of Proof: A Historical and Sociological Exploration. *IEEE Annals of the History of Computing* 17, no. 3: 7–29.

Madariaga, R., and K. B. Olsen. 2002. Earthquake Dynamics. In *International Handbook of Earthquake and Engineering Seismology, Part A,* ed. W. H. K. Lee, H. Kanamori, P. C. Jennings, and C. Kisslinger, 175–194. New York: Academic Press.

Masterman, M. 1970. The Nature of a Paradigm. In *Criticism and the Growth of Knowledge,* ed. I. Lakatos and A. Musgrave, pp. 59–89. Cambridge: Cambridge University Press.

McPherson, J. M. 1988. *Battle Cry of Freedom.* New York: Oxford University Press.

Misner, C. W., K. S. Thorne, and J. A. Wheeler. 1973. *Gravitation.* San Francisco: W. H. Freeman.

Monmonier, M. 1999. *Air Apparent: How Meteorologists Learned to Map, Predict and Dramatize Weather.* Chicago: University of Chicago Press.

Nidditch, P. H. 1957. *Introductory Formal Logic of Mathematics.* London: University Tutorial Press.

Nieminen, R. M. 2001. From Computer Crunching to Virtual Reality. In *Mathematics Unlimited—2001 and Beyond,* ed. B. Enquist and W. Schmid, 937–960. Berlin: Springer.

Pavelle, R., M. Rothstein, and J. Fitch. 1981. Computer Algebra. *Scientific American* (December): 136–153.

Pennisi, E. 2001. Genome Duplications: The Stuff of Evolution? *Science* 294: 2458–2460.

Popper, K. R. 1972. *The Logic of Scientific Discovery.* London: Hutchinson.

Press, W. H., B. P. Flannery, S. A. Teukolsky, and W. T. Vetterling. 1992. *Numerical Recipes in Fortran.* 2d ed. Cambridge: Cambridge University Press.

Robertson, D. S. 1998. *The New Renaissance: Computers and the Next Level of Civilization.* New York: Oxford University Press.

Robertson, D. S. 1999. Algorithmic Information Theory, Free Will and the Turing Test. *Complexity* 4, no. 3: 17–34.

Robertson, D. S. 2000. Goedel's Theorem, The Theory of Everything, and the Future of Science and Mathematics. *Complexity* 5, no. 5: 22–27.

Schroder, M. 1991. *Fractals, Chaos, Power Laws.* New York: W. H. Freeman.

Service, R. F. 2001. High-Speed Biologists Search for Gold in Proteins. *Science* 294: 2074–2077.

Solomonoff, R. J. 1964a. A Formal Theory of Inductive Inference, Part I. *Information and Control* 7: 1–22.

Solomonoff, R. J. 1964b. A Formal Theory of Inductive Inference, Part II. *Information and Control* 7: 224–254.

Song, X. 2002. The Earth's Core. In *International Handbook of Earthquake and Engineering Seismology, Part A,* ed. W. H. K. Lee, H. Kanamori, P. C. Jennings, and C. Kisslinger, 925–933. New York: Academic Press.

Stewart, I. 1992. *The Problems of Mathematics.* 2d ed. Oxford: Oxford University Press.

Szalay, A. S. 1999. The Sloan Digital Sky Survey. *Computing in Science and Engineering* 1, no. 2: 54–62.

Theocharis, T., and M. Psimopoulos. 1987. Where Science Has Gone Wrong. *Nature* 329: 595–598.

Tymoczko, T. 1998. The Four-Color Problem and Its Philosophical Significance. In *New Directions in the Philosophy of Mathematics,* ed. T. Tymoczko, 243–266. Princeton, N.J.: Princeton University Press.

Voet, D., and J. G. Voet. 1995. *Biochemistry.* 2d ed. New York: Wiley.

Weinberg, S. 1992. *Dreams of a Final Theory.* New York: Pantheon Books.

Weinberg, S. 2001. *Facing Up.* Cambridge: Harvard University Press.

Wilczek, F. 2002. A Piece of Magic. In *It Must Be Beautiful: Great Equations of Modern Science,* ed. G. Farmelo. London: Granta Books.

Wilson, E. O. 1998. *Consilience: The Unity of Knowledge.* New York: Alfred A. Knopf.

Wolfram, S. 2002. *A New Kind of Science.* Champaign, Ill.: Wolfram Media Inc.

Wordsworth, W. 1979. The Prelude, Book XI. In *The Norton Anthology of English Literature, Vol. 2,* ed. M. H. Abrams. New York: W. W. Norton.

INDEX